Discovering Patterns in Mathematics and Poetry

116

Internationale Forschungen zur Allgemeinen und Vergleichenden Literaturwissenschaft

herausgegeben von

Alberto Martino
(Universität Wien)

Redaktion: Ernst Grabovszki

Anschrift der Redaktion:
Institut für Vergleichende Literaturwissenschaft, Berggasse 11/5, A-1090 Wien

Discovering Patterns in Mathematics and Poetry

To Karla

Marcia Birken
and
Anne C. Coon

Enjoy the left and right-brain connections.
Marcia Birken and Anne C. Coon

Rodopi

Amsterdam - New York, NY 2008

Cover photo:
Marcia Birken

Cover design:
Pier Post

Le papier sur lequel le présent ouvrage est imprimé remplit les prescriptions de "ISO 9706:1994, Information et documentation - Papier pour documents - Prescriptions pour la permanence".

The paper on which this book is printed meets the requirements of " ISO 9706:1994, Information and documentation - Paper for documents - Requirements for permanence".

Die Reihe „Internationale Forschungen zur Allgemeinen und Vergleichenden Literaturwissenschaft" wird ab dem Jahr 2005 gemeinsam von Editions Rodopi, Amsterdam – New York und dem Weidler Buchverlag, Berlin herausgegeben. Die Veröffentlichungen in deutscher Sprache erscheinen im Weidler Buchverlag, alle anderen bei Editions Rodopi.

From 2005 onward, the series „Internationale Forschungen zur Allgemeinen und Vergleichenden Literaturwissenschaft" will appear as a joint publication by Editions Rodopi, Amsterdam – New York and Weidler Buchverlag, Berlin. The German editions will be published by Weidler Buchverlag, all other publications by Editions Rodopi.

ISBN: 978-90-420-2370-3

Printed in The Netherlands

Preface

We have been friends, colleagues, and cross-disciplinary collaborators for over twenty-five years. Our partnership began when we were new faculty teaching in separate departments of an academic support center at Rochester Institute of Technology (RIT). Together, we were asked to write the curriculum for a course in Creative Problem Solving that we subsequently team-taught for several years. We used analogy, logic, formal debates, and number games to draw students into thinking about the connections between concepts they were learning in technical fields and concepts in the humanities. The challenge of conducting research and developing curriculum across disciplinary lines was new to both of us. It forced us to consider our traditional fields of mathematics and English literature in new ways. To encourage our students, we actively sought for ourselves the connections we wanted them to see. We collected countless examples of problems, readings, and images that linked our fields. As we went on to establish our separate careers in the Colleges of Science and Liberal Arts at RIT, we also continued to pursue opportunities to work together. We created overlapping introductory courses in writing and math for mathematics majors and used early on-line technology to facilitate student collaborations outside the classroom. We staged formal debates on the role of technology in education and presented interdisciplinary curricular models to conferences of English professors, mathematicians, and engineers.

In 2000, we took our collaboration further institutionally when we made the commitment to create a course that would bring together upper level students of all majors to explore the links between mathematics and poetry. The course, "Analogy, Mathematics, and Poetry," (later called "Patterns in Poetry and Mathematics") attracted many students who had strong interests in both math and poetry (some had given up one or the other as they concentrated on their more narrowly-focused fields of study). There were also students who loved music, Rubik's Cubes, martial arts, science fiction, and quilting. Many were drawn to the course simply because the premise of an unexpected interdisciplinary connection intrigued them. Each time we taught the course, we took the investigation further, inspired by the interests and intellect of talented students, who were themselves agile interdisciplinary thinkers.

This book is in many ways the embodiment of our process of collaboration. Moving from concrete to increasingly abstract ideas, it is focused, just as our collaboration has come to be, on the patterns that may be found within our two disciplines and the patterns that cross or are shared by both. Some are easy to spot; others require time and reflection. It is our hope that the book will inspire its readers to look for connections that may not be obvious, to take chances with ideas and categories, and to seize opportunities to work across boundaries.

We have been fortunate that our collaboration has evolved within the company of many very supportive individuals. At RIT, we were encouraged in the development of team-taught courses and were recognized for the curricular strength of our interdisciplinary collaboration. The formulation, research, and early drafts of this book were helped enormously by Faculty Development Grants we both received from our respective colleges and by sabbaticals granted by the Institute. Our current and former deans — Robert Clark and Ian Gatley in the College of Science and Andrew Moore and Glenn J. Kist in the College of Liberal Arts — were generous in their support of our collaboration and the links it formed between the two colleges. Other strong advocates of our work have included RIT's former President, Albert J. Simone; RIT's Provost, Stanley McKenzie, from whom we received a joint Provost's Learning Innovations Grant; and many colleagues from throughout the Institute. The RIT students, from engineers to illustrators, photographers to computer scientists, helped us test, stretch, and refine our thinking, especially in the course that led to this book. Our collaborative approach to teaching was warmly received in meetings of the biennial workshop of the International Society for the Study of European Ideas, chaired by Vladimir Fomichov, Russian State Technological University, Moscow. For editorial assistance as we prepared the manuscript, we thank Ernst Grabovszki of Editions Rodopi and Rahul Mehta. We also wish to acknowledge that all of the mathematical and photographic images, except in a few instances where noted, are the original work of Marcia Birken, who, as Professor Emeritus, is now traveling widely and photographing patterns in nature. Finally, we are indebted to our husbands, Eric Birken and Craig J. Zicari, whose sustained enthusiasm for this project was instrumental to its completion, and to our children, Adam Birken, David Birken, and Sarah Cirocco Angulo, whose maturation and flourishing so closely paralleled that of our collaboration.

Marcia Birken
Professor Emeritus, School of Mathematical Sciences
College of Science, Rochester Institute of Technology

Anne C. Coon
Professor of English and Sr. Associate Dean
College of Liberal Arts, Rochester Institute of Technology

Rochester, New York November 2007

Table of Contents

Preface 5

Introduction 9

Chapter 1— Counting Patterns 15
1.1 Countable Patterns in Mathematics 15
Number Systems 16
Number Sequences 21
Infinity 24
Playing with Numbers 26
1.2 Patterns in Rhythm, Rhyme, and Ideas 31
Simple Patterns We Can Feel and Hear 31
More about Rhyme 36
Traditional Forms: Limerick and Ballad 38
Patterns in Ideas: Counting and Listing 41

Chapter 2 — Counting Patterns Take Form 45
2.1 How Patterns Take Form in Mathematics 45
Pascal's Triangle 45
The Fibonacci Sequence 52
The Golden Ratio or Golden Number, Φ 61
2.2 Traditional Poetic Forms: Patterns and Adaptations 66
Sestina 66
Sonnet 76
Villanelle 83

Chapter 3 — Patterns of Shape 87
3.1 Mathematical Shapes 87
Plane Figures 88
Spirals 93
3.2 Shaped Poetry 96
Labyrinths and Puzzles 96
Geometric Forms 97
Modern Pattern Poetry 100

3.3 Symmetry in Mathematics 106
Basic Symmetries 106
Tessellations and the Art of M. C. Escher 117
3.4 Symmetry in Poetry 123
Symmetry in Structure 123
Symmetry and Organic Form 128
Symmetry as Metaphor 130

Chapter 4 — Fractal Patterns 131
4.1 Basic Fractal Concepts 133
Fractals in Nature 138
4.2 A Brief History of Fractal Geometry 141
Mathematical Thought Before Mandelbrot 141
Julia Sets and The Mandelbrot Set 147
Chaos and Fractals 155
4.3 Fractals and Poetry 156
Fractal Images in Words 156
The Mandelbrot Set and the Language of Poetry 157
Poetic Dimensions of Self-similarity 160
Reading with Fractals in Mind 162
Fractals as a Link Between Mathematics and Poetry 167

Chapter 5 — Patterns for the Mind 169
5.1 The Mathematical Mind: Proof, Paradox, and Infinity 169
Mathematical Proof 169
Paradox 176
Infinity Revisited 178
5.2 Ideas that "tease us out of thought" 181
Considering Language 182
Figuring out the Figurative 183
Where Lies the Proof? 184
Poetry and Paradox 186
Poetic Concepts of the Infinite 189

Chapter 6 — Conclusion 193

Permissions Acknowledgements 195

Bibliography 199

Index 207

Introduction

This book is about patterns, the surprising, sometimes parallel patterns found in mathematics and poetry. But what are patterns? Patterns are models or plans. Patterns are recognizable and powerful arrangements of words, numbers, or figures. Patterns can be found in the structures of folktales and in the predictable steps that lead to the punch line of a joke. Patterns occur throughout the natural world, from the shape of an unfolding fiddlehead fern to the minute structure of the genome. Patterns are recurring symbols or images linked to create a poem, a symphony, or a mural. Patterns give form to equations and provide ways to classify and categorize arithmetic, algebra, and geometry. Patterns are created by the repeated occurrences of cycles, seasons, and tides. Patterns help us make predictions and make learning possible. Patterns are progressions and repetitions. Patterns can be seen and felt and heard: the shape of a snail's shell, the syncopated beat in a child's skipping, the shrill repeated notes of a cardinal's song. Patterns embody qualities of beauty and proportion in the human body or a snowflake. Patterns comfort and sustain us, in heartbeat and breath.

Human beings recognize and are attracted to patterns. When children first learn about their world, they begin to recognize and master its patterns. Later, as adults, we organize and express our ideas in patterns and rejoice to find complex patterns around us. Much of our creative expression, exploration, invention, and knowledge is based on patterns. Patterns are implicit and explicit in two of the most familiar, and sometimes most mysterious, human endeavors: mathematics and poetry.

When we began this book, we anticipated the question you may be asking now: "Why talk about mathematics and poetry together?" Good question! Wouldn't other pairs of disciplines have more in common: poetry and history, perhaps? Or mathematics and music? Why bring together two fields that seem to many people so very disparate? What could mathematics and poetry share, except that the mention of either one is sometimes enough to bring an uneasy chill into a conversation? For years, in our collaborative teaching, writing, and research, we have worked to identify and explore the connections between mathematics and poetry; we have discovered many connections, both simple and complex, and some not obvious at first glance.

One important connection we will explore here is the use of analogy in mathematics and poetry. Both fields use analogies — comparisons of all sorts — to explain things, to express unknown or unknowable concepts, and to teach. Consider for a moment the indelible visual image of one of the simplest analogies associated with math: the "pie" cut into halves and then quarters that helps children visualize fractions. Or think of the enduring association drawn by Robert Burns and countless other poets between love and a red rose. Both of these familiar analogical comparisons originated as representations of concepts that were too large or too abstract to be easily described. The use of analogical thinking may also be subtle and sophisticated. A poet might use a metaphor to capture the ambiguities of war, while a mathematician might explain a new theory by comparing it to simpler, known concepts. Throughout the following chapters, as we present patterns of all kinds, we will often return to the role analogy plays in creating or understanding those patterns.

Understanding analogy is just one aspect of a student's learning in mathematics and poetry. Let's consider for a moment how that learning takes place in a more general way. Almost simultaneously in their development, children learn to recognize both numbers and letters; they learn to count and to recite the alphabet. In some of a child's earliest learning, poetry and math are closely linked, as in the counting rhyme, "one, two, buckle my shoe," where simple concepts from both areas are playfully linked. But learning about poetry and mathematics soon diverges. When children enter school, their formal mathematical instruction begins, and continues for years, introducing increasingly difficult content and often relying on memorization and mastery of prescribed formulas. By the time they are in high school, students are mastering skills of computation, algebra, geometry, and trigonometry, although they are usually solving problems for which there only single correct answers.

Although much of what is taught in high school mathematics is considered fundamental to a good education, there is also a practical component to the curriculum. For example, understanding basic mathematical concepts makes it possible for a person to calculate interest rates on an investment, to multiply fractions when increasing a recipe, or to estimate the height of a building seen at a distance. Usually, only those students who go on to study mathematics in college have the opportunity to explore open-ended problems where abstract thinking is emphasized over formulaic methods of solution. Students who major in mathematics learn that mastery of basic skills and memorization of formulas are just the beginning. In order to understand the world in mathematical terms, they must begin to focus on abstract reasoning and analogous methodologies. No longer is it sufficient to learn the material for a single course in isolation; instead, the material in calculus lays a foundation for studying differential equations, both of which, in turn, prepare students to explore systems of equations in more theoretical ways.

The experience of learning about poetry is somewhat different. Long before they enter school, many children are charmed by reciting nursery rhymes. Young children learn that they can make poetry by saying words aloud and later by putting them together on paper. In school, students learn some of the technical terminology associated with poetry, much as they learn mathematical formulas. They read and discuss individual poems, frequently those that are famous or easily memorized. In middle or high school, young people may continue to write poetry on their own, using it to express their emotions or frustrations. Ironically, at the same time, these same students may be losing interest in poetry as an area of study, finding it hard to understand, too flowery, or boring. Unlike mathematical competence, the understanding and appreciation of poetry is much more difficult to measure, and generally seen as less critical to a person's education. Unless they are English majors, college students may read little more than a few poetry selections included in introductory literature classes. Nonetheless, many people do read and remember poetry long into adulthood. At times of joy or grief — at weddings and funerals and when they're in love — they turn to poetry for words that will describe their exultation or give them solace.

Despite the usefulness of mathematics as a life skill and the comfort poetry provides in times of emotional stress, the work of the poet and that of the mathematician are often regarded with skepticism. Outside the classroom, poetry and mathematics may be seen as inaccessible, their languages dismissed as cryptic, even forbidding to an outsider. In this book, we do not intend to make every reader into a poet or mathematician, but, instead, as we study the patterns of poetry and mathematics, we hope to "translate" their languages for a general audience. Likewise, as we make these languages more understandable, we hope that the patterns of each field will become open to appreciation by a wider audience.

Our interdisciplinary approach to writing this book creates some unique problems. For example, seemingly simple words such as *measure*, *rational*, or *line* mean different things to a poet and a mathematician. By explaining critical terminology, making connections and moving back and forth between the two disciplines, we hope to engage and challenge the reader, to offer the background necessary for understanding, and to highlight the creative possibilities shared by mathematics and poetry. As you read the book, you may find the following information helpful:

1. Although the chapters build in complexity, we recognize that a reader's interest and background are the best guide to approaching the book's material. If you want to jump right in and read about fractals, go ahead! The chapters are meant to be enjoyed together, but you may want to read them in an order that makes the most sense to you.

2. To link the discussions among chapters and to illustrate the different ways one may approach a poem, we sometimes discuss the same poem in more than one chapter. We hope that this recursive approach deepens your understanding and encourages you to ask questions of your own about patterns in the poem's ideas and structure.

3. Occasionally, we introduce an advanced mathematical explanation that makes a leap beyond the material being covered. These explanations will be boxed in the text, and you could omit them without losing the continuity of the material.

The writings of many scholars, teachers, poets, and mathematicians have been useful to us in our work although no single volume shares our specific focus. There are books about patterns in mathematics and mathematical patterns in nature, as well as books linking math with art, music, and architecture. Other books describe the structure and forms of poetry, compare poetry with music, and combine art or photography with poetry. Several poetry collections focus on scientific and mathematical themes. Some mathematical books make brief references to poetry and vice versa. Because we are approaching mathematics and poetry together, and because of our focus on both the structural and abstract qualities of patterns, this book is different.

In Chapter 1, "Counting Patterns," we explore the fundamental patterns that may be counted, felt, seen, and heard in mathematics and poetry. We demonstrate how mathematicians base much of their work on certain naturally-occurring patterns, and we trace the development of human understanding of number systems and number sequences. We also examine patterns in rhyme and meter that arise from breath and movement and from our ability to generate and recognize sound. In Chapter 2, "Counting Patterns Take Form," we move from Pascal's Triangle and the pleasing geometric pattern of the Golden Ratio to three selected traditional poetic forms, the sestina, sonnet, and villanelle. After reviewing the origins and organizing principles of these forms, we look at how poets have adapted and experimented with them, creating innovative work that still draws upon the traditional patterns. Chapter 3, "Patterns of Shape," is devoted to spirals, labyrinths, symmetrical shapes, and abstract shapes. Here, visual elements of math and poetry often come together in surprising ways. Chapter 4, "Fractal Patterns," begins with a general description of fractals, illustrated with examples of the abundance of fractal patterning found in the natural world. We then review the history of fractal geometry and explore some of the influences fractals have had on the reading and writing of poetry. In Chapter 5, "Patterns for the Mind," our focus is on the brain-teasing concepts of proof, paradox, and infinity. We look at how these concepts have been understood and represented in mathematics and then show how they have been interpreted imaginatively in the structure and ideas of poetry.

Discovering Patterns in Mathematics and Poetry is the result of an interdisciplinary collaboration sustained over twenty-five years. To borrow from the language of fractal geometry, our collaboration has undergone many iterations as our research, writing, and teaching interests have expanded and been refined. Throughout the various iterations, we have always believed that our two disciplines have a great deal to say to and learn from one another. For us, that dialogue begins with patterns.

Chapter 1 — Counting Patterns

1.1 Countable Patterns in Mathematics

To begin our exploration of mathematical patterns, we will first examine numbers, their arrangements in lists, and their development into number systems. By "number," we mean a symbol or word used to count, order, measure, or locate. Numbers are the essential building blocks of mathematics, the essential tool of arithmetic, as described by Carl Sandburg:

> Arithmetic is where numbers fly like pigeons in and out of your head.
> Arithmetic tells you how many you lose or win if you know how many you had before you lost or won.
> Arithmetic is seven eleven all good children go to heaven - or five six bundle of sticks.
> Arithmetic is numbers you squeeze from your head to your hand to your pencil to your paper till you get the answer.
> Arithmetic is where the answer is right and everything is nice and you can look out of the window and see the blue sky - or the answer is wrong and you have to start all over and try again and see how it comes out this time.
> If you take a number and double it and double it again and then double it a few more times, the number gets bigger and bigger and goes higher and higher and only arithmetic can tell you what the number is when you decide to quit doubling.
> Arithmetic is where you have to multiply - and you carry the multiplication table in your head and hope you won't lose it.
> If you have two animal crackers, one good and one bad, and you eat one and a striped zebra with streaks all over him eats the other, how many animal crackers will you have if somebody offers you five six seven and you say No no no and you say Nay nay nay and you say Nix nix nix?
> If you ask your mother for one fried egg for breakfast and she gives you two fried eggs and you eat both of them, who is better in arithmetic, you or your mother?
>
> (Sandburg 655-56)

Numbers have provided humans with a mechanism to group things, organize their world, create symbols, play games, and keep time. The earliest number patterns were probably first observed in physical attributes and in nature: the appendages of fingers and toes; the pairings of eyes, ears, hands, and feet; the cycles of the sun and moon; the groupings of stars in the sky; and the recurrence of changing seasons. Regular patterns were too common to ignore. Each human had the same number of fingers, even if the numbers 5 and 10 had not yet been named. The sun, moon, stars, and seasons returned in fixed, repeating cycles, even if clocks, star charts, and calendars had not been invented. The earliest recorded numbering began very simply with marks on the ground, a stick, or cave wall. Ancient languages indicate that the first spoken concepts of counting were limited to very few words, such as "one, two, many."

As the hunter-gatherers settled in groups and began to cultivate crops, they developed ways to count large quantities, to place items in groups, to compare size, and to record trades. The patterns of numbers expanded to include the more abstract concepts of "larger than" and "more of the same quantity." The early counting pattern of "one, two, many" expanded to what we know today as the *Counting* or *Natural Numbers*.

Number Systems

Understanding the development of our modern number system will help us grasp more complex number patterns. What follows is not meant to be a comprehensive history of the evolution of numbers, but rather a brief outline of the relationship between the various types of numbers we use today.

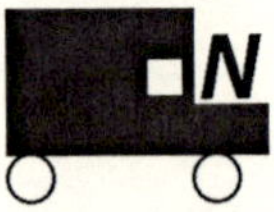

The *Natural Numbers* consist of the numbers we use in counting. Usually denoted by *N*, the Natural Numbers can be written down as a list. The three dots at the end of the list indicate that the numbers go on forever. These numbers are, of course, our simple counting numbers.

$$1, 2, 3, 4, 5, 6, 7, 8, \ldots$$

Our ancestors made great progress in abstract thinking when they went from simple counting to incorporating the idea of nothing or zero into the number arrangement. It is unclear when or even where the concept of zero first appeared, but there is some evidence of such thought in ancient Babylonian,

Greek, and Hindu mathematics. The status of zero as a number, rather than an idea, remained shrouded in secrecy and mystery well into more modern times. Even today, zero is different from the other numbers. No number, not even zero itself, can be divided by zero! When we add zero to the list of Natural Numbers, the new list is called the *Whole Numbers*:

$$0, 1, 2, 3, 4, 5, 6, 7, 8, \ldots$$

Another conceptual leap occurred with the idea that *negative* numbers were necessary to complete this part of the number system. Although there seems to be some use of negative numbers in ancient Chinese and Indian writings, the idea of negative numbers was strongly resisted in most early cultures. One could easily represent 3 or 17 by placing the corresponding number of objects in a pile, but how could -3 be physically represented? If one tried to take away 13 beads from a pile of 10 beads, the number left was none. Early notions of negative numbers were often related to something evil or unlucky. Until humans could think about numbers as abstractions, rather than symbols of physical objects, people did not accept the idea that negative numbers existed. In time, the taboo associated with negative numbers was overcome when the negative sign was associated with something concrete, such as direction. For instance, -3 may be thought of as a representation for three units to the left of some fixed point, while +3 corresponds to a location three units to the right of that point.

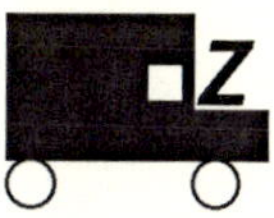

The *Integers* is the name given to the number system that includes both positive and negative counting numbers, as well as zero. The Integers, represented by *Z*, continue infinitely in both the positive and negative directions. One can think of zero as the "balancing point" of the list below, with half the numbers to its right and the other half to its left.

$$\ldots -4, -3, -2, -1, 0, 1, 2, 3, 4, \ldots$$

Although integers made it possible to count forward and backward, there is still much more to the story of numbers. Interactions between people required more complicated tasks, such as dividing and measuring, leading to the development of fractions.

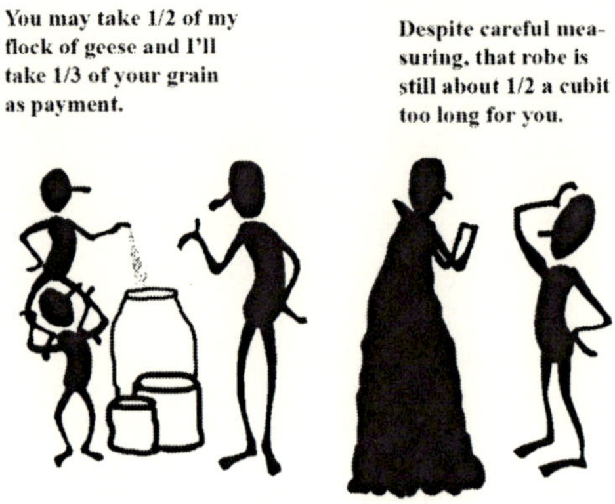

Figure 1.1

The *Rational Numbers,* usually denoted by *Q,* are all numbers that can be expressed as the ratio (or quotient) of two integers, such as $\frac{3}{5}$ and $\frac{-617}{3}$. The Integers are included within the Rational Numbers, since an integer, 5 for example, can be expressed as $\frac{5}{1}$, $\frac{10}{2}$, etc. Even some decimal numbers are Rational Numbers, namely the finite decimal numbers, such as 2.35, and the infinite repeating decimal numbers, such as $0.333333\overline{3}$ and $25.23712712\overline{712}$. In both examples the bar is placed over the number or group of numbers that repeats forever. Most of us are familiar with the fact that $0.333333\overline{3}$ equals $\frac{1}{3}$.

> Why is $0.333333\overline{3}$ equal to the rational number $\frac{1}{3}$?
>
> Let $N = 0.33333333333333\overline{3}$
>
> Then $10N = 3.33333333333333\overline{3}$
>
> Subtract N from 10N to eliminate the infinite decimals.
>
> $$\begin{array}{rcl} 10N &=& 3.33333333333333\overline{3} \\ -N &=& 0.33333333333333\overline{3} \\ \hline 9N &=& 3 \end{array}$$
>
> Therefore $N = \frac{3}{9} = \frac{1}{3}$.

The diagram in Figure 1.2 shows the relationship between the systems we've described so far.

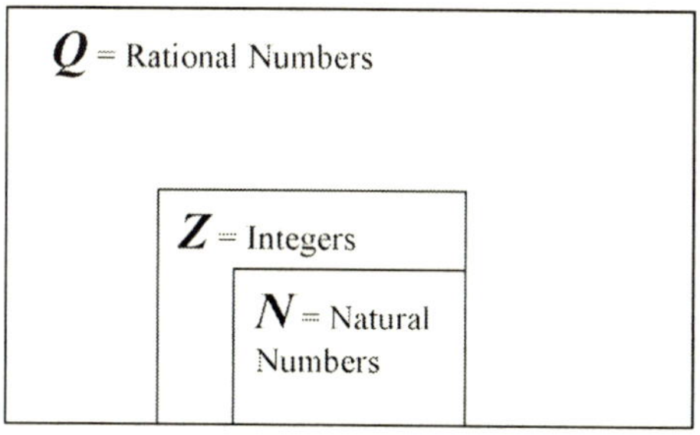

Figure 1.2

But the number story does not end here. The ancient Greeks, as well as other early civilizations, discovered that some measurements could not be represented by the Rational Numbers. About 500 years B.C.E., the Pythagoreans (followers of Pythagoras' school of mathematics and philosophy) proved that *Irrational Numbers* existed. The Irrationals, such as $\sqrt{2}$ and π, are numbers that cannot be expressed as the ratio of two integers. Their existence caused great consternation, even horror, among the Pythagoreans whose philosophy was based on the harmony of proportion represented by the ratio of integers. Despite the difficulties this new idea presented, resulting in legends of secrecy and suppression, the Irrational Numbers forced mathematicians to think about numbers in a more theoretical way.

The shape in Figure 1.3 is a square whose sides a and b are each one unit in length. The length of the diagonal c gives us a physical representation of an Irrational Number.

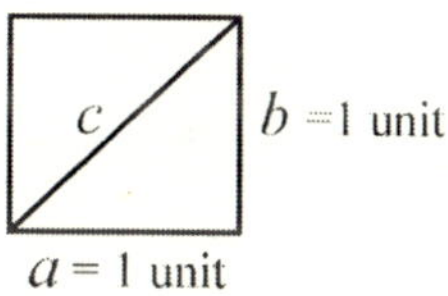

Figure 1.3

Using the Pythagorean Theorem, $c^2 = a^2 + b^2$, we can see that the length of c is $\sqrt{2}$ units. Unlike the Rational Number $\sqrt{4} = 2 = \frac{2}{1}$, the Irrational Number $\sqrt{2}$ cannot be expressed as the ratio of two integers. Irrational Numbers are infinite, non-repeating, decimal numbers, so the $\sqrt{2}$ would be represented as 1.414213. . .

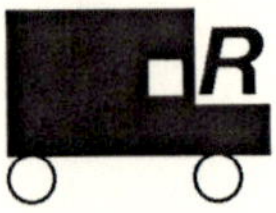

The Rationals and the Irrationals together make up the *Real Numbers*, denoted by *R*. These are the numbers used to solve math and physics problems in high school and college classes. Real Numbers measure continuous quantities like time, length, velocity, and temperature, although we often use decimal approximations to express them.

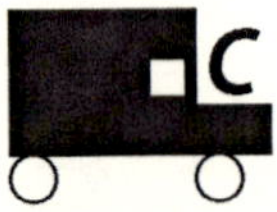

It wasn't until the sixteenth century that the last piece of this system was defined formally: the Complex Numbers, denoted by *C*. These are numbers that can be expressed in the form $a+bi$, where i stands for the imaginary number $\sqrt{-1}$. This imaginary number is a solution to the simple equation $x^2+1=0$. Some examples of Complex Numbers are $3+2i$, $-7i$, and $23+0i$ (which is the Real Number 23). Thus, the Real Numbers are contained within the Complex Numbers.

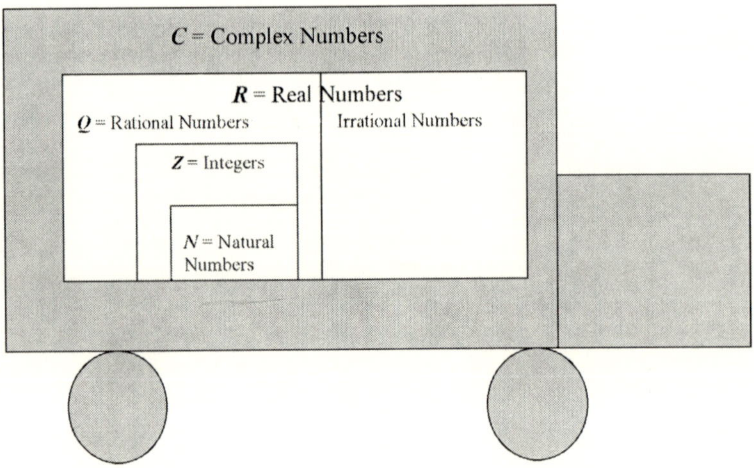

Figure 1.4

Figure 1.4 adds the Irrational, Real, and Complex Numbers to our earlier drawing (Figure 1.2) and shows the interrelationship of the various number systems we have described. Now that our brief overview is complete, we can begin describing and manipulating number patterns.

Number Sequences

The mathematical term *sequence* means an infinite list or string of numbers. For instance, the Natural Numbers and Whole Numbers are both sequences. The individual numbers in the list are called the *terms* of the sequence. Order is important when writing out a sequence, so we often call the first term a_1, the second term a_2, and so on. A general term of the sequence is referred to as the *nth term*, or a_n. Since the first term of the Natural Numbers is 1, $a_1 = 1$, $a_2 = 2$, and $a_n = n$ for this sequence. In contrast, the sequence of Whole Numbers begins with 0, so $a_1 = 0$, $a_2 = 1$, and $a_n = n - 1$.

If we can find a formula that represents every term in the sequence, then we can create a shorthand, such as $\{a_n\}_{n=1}^{n=\infty}$, to use in place of writing the long string of terms. In this notation the brackets around a_n indicate there is a sequence or listing of terms; a_n represents the nth term of that listing; and the two values of *n* on the right indicate that *n* starts at 1 ($n = 1$) and goes on forever ($n = \infty$). For example $\{n\}_{n=1}^{n=\infty}$ is the familiar sequence of Natural Numbers 1, 2, 3, 4, 5, . . ., while $\{n-1\}_{n=1}^{n=\infty}$ is the sequence of Whole Numbers 0, 1, 2, 3, 4, . . .

Some of the best-known sequences are formed by counting in fixed intervals. If we start at the number 1 and count every other number, we find the sequence of odd positive integers 1, 3, 5, 7, . . . Since $a_1 = 1$, $a_2 = 3$, $a_3 = 5$, we can say the *nth* term of the odd positive numbers is $a_n = 2n - 1$ and the odd positive numbers can be represented by $\{2n-1\}_{n=1}^{n=\infty}$.

A listing of the odd positive integers shows that we can make a correspondence between those integers and the Natural Numbers, as seen in Figure 1.5 below. This list should assist the reader in visualizing this abstract idea.

Sequence Name	$\{a_n\}$	a_1	a_2	a_3	a_4	a_5	a_6	...
Natural Numbers	$\{n\}$	1	2	3	4	5	6	...
Odd Positive Integers	$\{2n-1\}$	1	3	5	7	9	11	...

Figure 1.5

If we do exactly the same thing, but start at 2 and count every other number, we find the even positive integers. These are the numbers that can be expressed in the form $2n$, where n is a Natural Number. Figure 1.6 shows the correspondence between the even positive integers and the Natural Numbers.

Sequence Name	$\{a_n\}$	a_1	a_2	a_3	a_4	a_5	a_6	...
Natural Numbers	$\{n\}$	1	2	3	4	5	6	...
Even Positive Integers	$\{2n\}$	2	4	6	7	8	10	...

Figure 1.6

Mathematicians enjoy observing the patterns in these sequences, as well as finding symbolic ways to express them, such as $\{2n-1\}$. Determining the shorthand for a_n is a large part of the fun in pattern recognition. For instance, if $\{2n-1\}_{n=1}^{n=\infty}$ produces the odd positive integers, what does $\{2n+1\}_{n=1}^{n=\infty}$ produce? Starting at $n = 1$, we get the sequence of numbers 3, 5, 7, 9, . . ., or the odd positive integers beginning with 3. So another way to produce ALL the odd positive integers would be to use $\{2n+1\}_{n=0}^{n=\infty}$, that is, let n begin with zero. Then we could find a correspondence between the Whole Numbers and the odd positive integers, as shown in Figure 1.7.

Sequence Name	$\{a_n\}$	a_1	a_2	a_3	a_4	a_5	a_6	...
Whole Numbers	$\{n\}$	0	1	2	3	4	5	...
Odd Positive Integers	$\{2n+1\}$	1	3	5	7	9	11	...

Figure 1.7

What would the sequence $\left\{\frac{3n}{n^2}\right\}_{n=1}^{n=\infty}$ look like? The first term, a_1, is $\frac{3\cdot 1}{1^2}$ or 3; the second term is $\frac{3\cdot 2}{2^2}=\frac{6}{4}$ or $\frac{3}{2}$; and the third term is $\frac{3\cdot 3}{3^2}=\frac{9}{9}$ or 1. If we listed these first three numbers in simplified form as 3, $\frac{3}{2}$, and 1, it would be difficult to tell the pattern that is forming our sequence. The listing for individual terms in the sequence $\left\{\frac{3n}{n^2}\right\}_{n=1}^{n=\infty}$ is given in Figure 1.8.

Sequence Name	$\{a_n\}$	a_1	a_2	a_3	a_4	a_5	a_6	...
Natural Numbers	$\{n\}$	1	2	3	4	5	6	...
New Sequence	$\left\{\frac{3n}{n^2}\right\}$	3	$\frac{3}{2}$	1	$\frac{3}{4}$	$\frac{3}{5}$	$\frac{3}{6}$	...

Figure 1.8

If we look carefully at the listing, we can see a second way to represent this sequence. The sequence produced by $\left\{\frac{3n}{n^2}\right\}_{n=1}^{n=\infty}$ is exactly the same as that produced by $\left\{\frac{3}{n}\right\}_{n=1}^{n=\infty}$. If this type of mathematical pattern intrigues you, try unraveling the patterns in the boxed text.

1. What are the first 5 terms of the sequence $\left\{\frac{n+1}{3n-1}\right\}_{n=0}^{n=\infty}$?

   One answer is $\frac{1}{-1}, \frac{2}{2}, \frac{3}{5}, \frac{4}{8}, \frac{5}{11}, \ldots$

   Another is -1, 1, $\frac{3}{5}, \frac{1}{2}, \frac{5}{11}, \ldots$

2. Write the sequence $\frac{3}{4}, \frac{5}{4}, \frac{7}{4}, \frac{9}{4}, \frac{11}{4}, \ldots$ in the form $\{a_n\}$.

   One answer is $\left\{\frac{2n+3}{4}\right\}_{n=0}^{n=\infty}$.

   Another is $\left\{\frac{2n+1}{4}\right\}_{n=1}^{n=\infty}$.

Infinity

We have used three dots at the end of each sequence to indicate that the pattern continues infinitely, but what exactly does "infinitely" mean? The topic of infinity is so important to our discussion of math and poetry that we will return to it in Chapter 5 when we discuss patterns of the mind. For the moment, let's discuss how the word *infinity* is used in mathematics. Infinity is not a number, not an object, not a formula. Infinity is a concept. It is not one single concept, but rather there are classes of infinity. These classes are very different from each other, but can be compared to each other. There are smaller and larger infinities within this hierarchy. Don't be surprised if you have difficulty grasping infinity on your first reading of this section. Although the ancient Greek mathematicians finally accepted Irrational Numbers such as π (an infinite, non-repeating, decimal number), they were not ready to fully grasp the concept of infinity. We see this in the paradoxes of Zeno of Elea who was born around 490 BCE. In one paradox, Zeno claims motion is impossible, since to cover the distance from one point to another we must first cover half the distance, and then cover half the remaining distance, and so on *ad infinitum*. Thus we would never reach the final destination, since there would always be half the final distance to cover. Zeno could not resolve this paradox without the later ideas of convergence and limits.

Infinity was a troubling concept in ancient times, but it could also be a comforting one. Often infinity was associated with God, and so the "unknowable" was tied into a reassuring belief system. In Genesis 13:16, we have the promise from God to Abram, "I will make your descendants as the dust of the earth; so that if one can count the dust of the earth, your descendants also can be counted." Again, in Genesis 15:5, God says to Abram, "Look toward heaven, and number the stars, if you are able to number them [. . .] so shall your descendants be." Finally, in Genesis 22:17, God promises, "I will indeed bless you, and I will multiply your descendants as the stars of heaven and as the sand which is on the seashore" (*The Oxford Annotated Bible*).

It wasn't until the sixteenth century that infinity again emerged as an important topic in mathematics. During the intervening years, treatises on infinity were less related to science and more a discussion of theology. Infinity was simultaneously revered, when associated with a higher being, and feared, when associated with ultimate evil. In 1592 the French mathematician François Viète showed a formula for finding an approximation to π that involved only the number 2. For the first time, a formula showed an explicit mathematical equation involving an infinite process.

$$\frac{2}{\pi} = \frac{\sqrt{2}}{2} \cdot \frac{\sqrt{2+\sqrt{2}}}{2} \cdot \frac{\sqrt{2+\sqrt{2+\sqrt{2}}}}{2} \cdots$$

(Maor 10)

Now infinity was "legitimized" and could be accepted by mathematicians. By the late 1600s, a huge leap in mathematical thinking would take place when Gottfried Leibnitz in Germany and Isaac Newton in England simultaneously developed their theories of calculus, an area of mathematics that relies heavily on the "infinitesimal" or the infinitely small.

The next major leap in mathematical thinking about infinity occurred in the nineteenth century when the mathematician Georg Cantor published a series of works that showed there is not one single concept of the infinite, but rather there are hierarchies of infinity which are different from each other. These different classes of infinity can actually be compared in size, with some infinities greater than others! Cantor's work was radical and questioned by many contemporaries.

Cantor showed that there are two types of infinite sets, those that can be placed in a correspondence with the Natural Numbers, called *countably infinite* sets, and those that cannot be placed in such a correspondence, called *uncountably infinite* sets. The Natural Numbers themselves, as well as the Integers and the Rational Numbers are countably infinite, while the Irrational Numbers, and hence the Real Numbers, are so "dense" that it is impossible to list them in a correspondence with the Natural Numbers. The Irrationals and the Reals are uncountably infinite. In Chapter 5 we will return to the complex, and often impenetrable, concept of uncountable infinity, but for the moment let's concentrate on some intriguing notions from countable infinity.

When discussing the size of a finite set of numbers, it is clear that two finite sets of numbers that can be placed in correspondence with each other would have the same number of elements, and thus be of the same size. Infinite sets are much stranger. Clearly, the set of positive even integers contains half as many numbers as the set of all positive integers, but they are the same size! That is because all countably infinite sets have the same size, which Cantor called $\aleph_0$ (aleph naught), taken from the first letter of the Hebrew alphabet. Also, it seems that the entire set of Integers should be twice as large as the positive Integers, but they are also the same size. The table in Figure 1.9 shows some countably infinite sets placed in correspondence with the Natural Numbers. Each set has size $\aleph_0$.

Natural Numbers	1	2	3	4	5	6	. . .
Integers	0	-1	1	-2	2	-3	. . .
Even Positive Integers	2	4	6	8	4	10	. . .
Odd Negative Integers	-1	-3	-5	-7	-9	-11	. . .
Multiples of 5	5	10	15	20	25	30	. . .
Squares of the Natural Numbers	1	4	9	16	25	36	. . .

Figure 1.9

One baffling property of countably infinite sets is that an infinite subset of a countably infinite set has the same size as the entire set. Such conundrums perplexed, intrigued, and even frightened mathematicians, but they also fired the imagination of poets. We will close this section on infinity with a verse written in Latin in *Ars Conjectandi* by the seventeenth-century Swiss mathematician Jacob (Jacques) Bernoulli.

Even as the finite encloses an infinite series
And in the unlimited limits appear,
So the soul of immensity dwells in minutia
And in the narrowest limits no limit in here.
What joy to discern the minute in infinity!
The vast to perceive in the small, what divinity!

(Maor, epigraph 25)

Playing with Numbers

From ancient times people have enjoyed playing with numbers in games, puzzles, and word problems. Often these mathematical recreations ask us to find or follow patterns in the numbers. Most people have heard the following old math riddle, given in the form of a poem:

As I was going to St. Ives,
I met a man with seven wives,

Each wife had seven sacks,
Each sack had seven cats,
Each cat had seven kits.
Kits, cats, sacks, and wives,
How many were going to St. Ives?

(*Mother Goose* 73)

Although one can diligently compute the total of the man and his 7 wives, sacks, cats, and kits, in fact the riddle expects an answer of only one person going to St. Ives, the "I" of the narrator. The man and his wives were coming from St. Ives when met by the narrator. One might argue that there is enough ambiguity to assume that the narrator overtook the man, his wives and assorted cats and kits, all heading to St. Ives, but that was not the intent of this eighteenth century puzzle-poem. Similar riddles and puzzles, offering challenging patterns of numbers that may or may not be useful to solving the problem, appear in texts across time periods and cultures.

A different form of number play is found in "magic" games in which one person asks a second person to think of a number. After a series of deliberately confusing computations, the first person mysteriously guesses the correct answer to the computations. Here is such an example:

Choose a number.
Triple it and add 20.
Double your answer.
Subtract 25.
Divide your answer by 3.
Subtract twice the number you originally chose.
Your answer is 5!

Of course, the problem is set up so that the answer is always 5, no matter what the initial number choice. The many computation steps are a diversion to confuse the person choosing the original number, but a little algebra reveals that no matter what number one begins with, the order of operations gives an answer of 5.

STEP	ANSWER
1. Choose x as your initial number	x
2. Triple it and add 20	3x + 20
3. Double your answer	6x + 40
4. Subtract 25	6x + 15
5. Divide your answer by 3	2x + 5
6. Subtract twice the number you chose	5

A different math game asks you to choose between two patterns of numerical computation. Perhaps your job offers two possible compensation plans for the first year of employment. Under the first plan you would earn a fixed salary of \$2000 per week for 52 weeks. Under the second plan you are offered 1 cent the first week, 2 cents the second week, 4 cents the third week, 8 cents the fourth week, and so on. Which plan is better at the end of one year? The unknowing employee might choose the first plan, but in fact the second pattern offers far, far more money.[1]

The magic square is our final example of playing with number patterns. This ancient arrangement of numbers in rows and columns appears in manuscripts, artwork, and books. It has even been worn as a talisman. What makes a square array of numbers into a magic square? In a *normal magic square of order n*, the Natural Numbers from 1 to n^2 each appear once in the square. These numbers are arranged so that the sum of any row, column, or main diagonal gives the same answer. There is only one normal magic square of order 3 and it is shown below. Notice that there are three rows, three columns, and each of the numbers from 1 to 3^2 (or 9) appears once in the square. Each of the three rows, three columns, and two main diagonals adds up to 15.

8	1	6
3	5	7
4	9	2

Figure 1.10

When the size of the magic square increases, the number of possible arrangements increases as well. A normal magic square of order 4 appears in the upper right corner of Albrecht Dürer's engraving "Melancholia I," a German work from the early 1500s.

[1] The first plan yields \$2000 x 52 or \$104,000 at the end of one year. The second plan offers 2^0 cents the first week, 2^1 cent the second week, 2^2 cents the third week, up to 2^{51} cents in week 52. At the end of 52 weeks, plan two yields a total of $S = 2^0 + 2^1 + 2^2 + 2^3 + ... + 2^{51}$. Multiply S by 2 and perform the subtraction $2S - S$ to discover $S = 2^{52} - 2^0$ or a salary total of \$45,035,996,273,704.95!

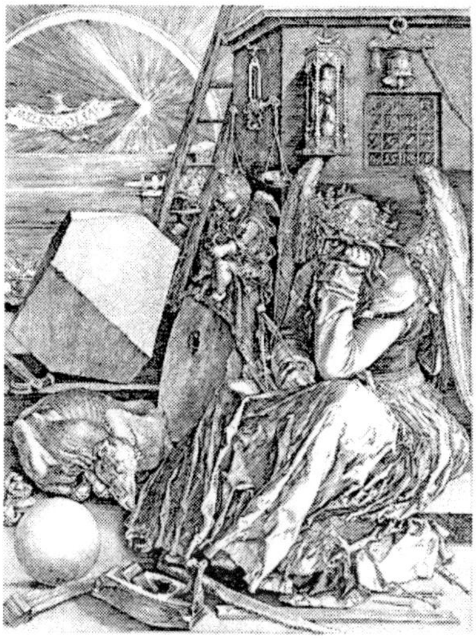

Figure 1.11

As with every normal magic square, each row, column, and main diagonal adds to the same number – here 34, but there is more "magic" to Dürer's pattern. The four interior cells add to 34, as do the four corner cells. Also adding the two middle cells in the first row to the two middle cells in the fourth row results in a sum of 34. Likewise, the two middle cells in the first column and fourth column add to this same sum. Figure 1.12 gives a clearer view of this famous square.

16	3	2	13
5	10	11	8
9	6	7	12
4	15	14	1

Figure 1.12

There are additional patterns hidden in this square of order 4, as well as other types of magic squares and cubes.[2] A far older and more complex magic square of order 6, shown in Figure 1.13, was photographed by one of the authors during a recent visit to China. This iron plate is inscribed with the numbers 1 through 36, arranged so that each number appears only once, and every row, column, and the two diagonals sum to 111. It dates from the Yuan Dynasty (1271-1368) and was found buried under the cornerstone of the home of Prince Anxi near Xi'an. Such relics were treated as sacred objects in ancient China and were buried to protect the home and exorcise evil spirits. Figures 1.13a and 1.13b show both the original iron plate, on display in Shanghai Museum, as well as a modern translation of the numbers.

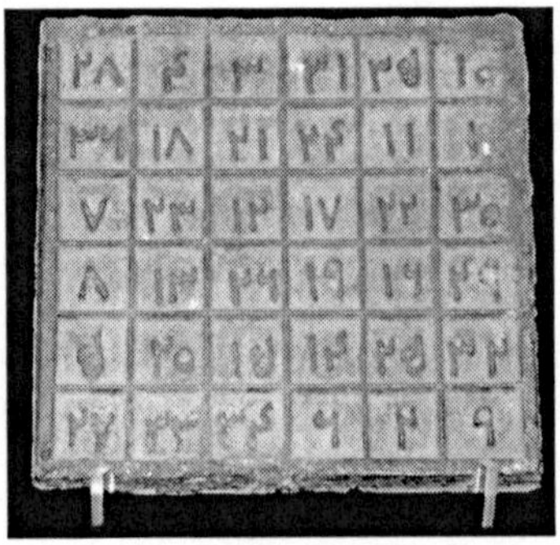

Figure 1.13a

28	4	3	31	35	10
36	18	21	24	11	1
7	23	12	17	22	30
8	13	26	19	16	29
5	20	15	14	25	32
27	33	34	6	2	9

Figure 1.13b

Once we start playing with these number patterns, it is tempting to get sidetracked. But let's return to the main focus of this chapter — finding the basic patterns of math and poetry and seeing how they intersect. Before exploring three interrelated patterns that are quite famous in mathematical circles, we will examine the basic, countable patterns of poetry.

[2] An excellent explanation of number squares, including further discussion of Dürer's Magic Square, is found in Graham Flegg's work *Numbers, Their History and Meaning*. New York: Schocken, 1983, 230 -236.

1.2 Patterns in Rhythm, Rhyme, and Ideas

Simple Patterns We Can Feel and Hear

In poetry, some of the most familiar patterns are not only counted, but also *felt* and *heard*. Those patterns have intrinsic structures, sometimes echoing the rhythm of the human heartbeat, the syncopation of a skipped step, or the steady pace of a strong runner. Long before a child is introduced to the formal study of poetry or mathematics, he or she may be charmed by the patterns in this poem:

> Hickory, dickory, dock!
> The mouse ran up the clock;
> The clock struck one,
> The mouse ran down,
> Hickory, dickory, dock!

(*Mother Goose* 87)

Nursery rhymes teach children to recognize patterns of sound and rhythm. When the lines of "Hickory, Dickory, Dock" are recited aloud, almost inevitably by the time the speaker comes to the end of the second line, the listener knows what to expect. Even a very young child will respond to the repeated, emphasized sounds of "dock" and "clock." The poem is also easy to memorize because of its combination of rhyme and rhythm. The added element of playfulness ("The mouse ran up the clock") brings something else to this poem: It makes a joke of the usual story of a mouse creating a scare; this mouse is frightened by the innocent tolling of the clock and runs back down.

Reading this poem to a child would be different from reading a bedtime story. While this nursery rhyme does in fact tell a tightly compressed story with its own surprise ending, the use of rhyme and rhythm adds an element that is often missing in stories. Despite its simplicity, "Hickory, Dickory Dock" brings together a complete system of sound, beat, and movement. It would be hard to read this poem — and other nursery rhymes, as well — without involving the head and the eyes, even arms, hands, and fingers, in the physical movements of nodding, clapping, or finger-snapping. This powerful kinesthetic pattern of beats, created by the ways words are said aloud, gives "Hickory, Dickory Dock" its rhythm. When the rhythm of a poem is felt in regular, consistent patterns, as it is here, we refer to it as *meter*. This is obviously an instance where a key term, *meter*, means something entirely different in poetry from what it means in mathematics.

The meter of poetry is born of physical elements: breath, lips, teeth, tongue, and vocal cords. Meter may be felt throughout the body. Try saying these words aloud:

Hickory, dickory, dock!

The combination of letters helps create the meter in this poem, as the speaker emphasizes the words or syllables that end with the repeated "ck" sounds. The emphasized syllables in "hick..." and "dick..." are both followed by two lighter, unstressed syllables "or-y." If we capitalize the stressed syllables, we can see this more clearly:

HICKory, DICKory, DOCK!
The MOUSE ran up the CLOCK;

This poem is an example of *accentual verse,* where each line or a group of lines in a poem has the same number of stressed beats. In this case, there are 3 stressed beats in lines 1, 2, and 5. And there are 2 stressed beats in lines 3 and 4, so that the metrical pattern changes in the lines where the complication of the story occurs.

A mathematician might be pleased to see that the next nursery rhyme introduces children to the familiar counting pattern of the Natural Numbers, but most people who recite or hear this poem are probably more aware of its lilting patterns of both rhythm and rhyme. Try reading it aloud to hear the patterns for yourself.

ONE, TWO, BUCKle my SHOE;
THREE, FOUR, SHUT the DOOR;
FIVE, SIX, PICK up STICKS;
SEVen, EIGHT, LAY them STRAIGHT;
NINE, TEN, a GOOD fat HEN;
eLEVen, TWELVE, DIG and DELVE;
THIRteen, FOURteen, MAIDS are COURting;
FIFteen, SIXteen, MAIDS in the KITchen;
SEVenteen, EIGHteen, MAIDS are WAIting;
NINEteen, TWENty, MY plate's EMPty.

(*Mother Goose* 63)

This poem is another example of accentual verse. In this case, there are 4 stressed beats in each line, regardless of the number of syllables. In fact, the number of syllables per line in this poem ranges from 5 to 9. We have used capital letters to indicate the stressed syllables in each line. The remaining syllables do not receive the same emphatic stress, and you should be able to see and hear how they vary in number from line to line. Accentual verse is commonly found in nursery rhymes, where the combination of strong stresses and simplicity makes the metrical pattern very easy to recognize, whether one is listening to the poem being read aloud or simply feeling the beats when reading it silently.

Another essential pattern that we recognize in poetry is the use of sound. In "One, two, buckle my shoe," the pattern is created when a word in the middle of the line has the same sound as, or rhymes with, the last word in the line. The middle word is also the second new counting number named in each line, so we hear the following pairs of rhymed words:

two . . .shoe,
four . . .door
six . . .sticks, and so on.

In some nursery rhymes, the repeated sounds occur at the ends of the lines. The arrangement of the pattern of end rhyme can vary a great deal. In this poem, we can hear the end rhyme in pairs of successive lines:

Molly, my sister, and I fell out,
And what do you think it was all about?
She loved coffee and I loved tea,
And that was the reason we couldn't agree.

(*Mother Goose* 29)

In other cases, the same rhyming sound is heard in every line of the poem.

Little Blue Ben, who lives in the glen,
Keeps a blue cat and one blue hen,
Which lays of blue eggs a score and ten;
Where shall I find the little Blue Ben?

(*Mother Goose* 119)

As we go on to consider more complex poems throughout this book, we will find that patterns of meter and sound often come together in powerful ways.

Meter is one of the most critical patterns in poetry. In addition to accentual verse, described above, the other metrical forms in English are the relatively uncommon *syllabic verse* where the number of syllables, not accents or stresses, is counted in the line; and *accentual-syllabic verse*, where the pattern of stressed or accented beats is measured in combination with the number of syllables in the line. Accentual-syllabic verse is the most common pattern found in metrical poetry written in English.

It's important to note here that not all poetry in English has easily recognized metrical patterns. Indeed, contemporary "free verse" often looks and sounds quite different from the traditional forms of poetry we will be introducing in this chapter. Even in free verse, however, where the writer may be turning away from formal patterns of rhyme and meter, we will often find patterns in the way he or she uses words, both for their sound and their rhythmic or accented values. Before looking at experimentation or variations in meter and rhyme, let's consider the basic elements of what is known as *prosody*, the study of the metrical structure of verse.

There is quite an array of terms used to describe different aspects of meter, but it's not necessary to know all of the terminology in order to understand, enjoy, or even write poetry. Nonetheless, the terminology is useful when one wants to be precise in describing the metrical patterns of a poem. As with any specialized language, having words to describe meter may also help us see or hear more within a poem and help us understand more about the craft that went into a poem's creation. For our purposes, we will introduce some of the most commonly used terms and a simple system of notation.

The terminology traditionally used to describe the meter of accentual-syllabic verse begins with the concept of *feet*. Each foot is a measure of rhythm that contains 1 stressed syllable and 1 or more unstressed syllables. There are four basic types of metrical feet found in English poetry: the *iamb*, the *trochee*, the *dactyl*, and the *anapest*. One common system for indicating which syllables are stressed (or accented), and which are unstressed (or unaccented) uses the acute accent mark (´) and the breve (˘), placing the symbol over the respective syllable, as shown in these examples.

Iamb: an unaccented syllable followed by an accented syllable ˘ ´
Example: "like THAT"

Trochee: an accented syllable followed by an unaccented syllable ´ ˘
Example: "brightly"

Dactyl: A single accented syllable followed by two unaccented ´ ˘ ˘
Example: "wittily"

Anapest: Two unaccented syllables followed by an accented one ˘ ˘ ´
Example: "as we find"

The following mnemonic verse, published by Marjorie Boulton fifty years ago, uses the qualities of each of the metrical feet to help students remember how to tell them apart.

˘ ´ ˘ ´ ˘ ´ ˘ ´

Iambic feet are firm and flat

˘ ´ ˘ ´ ˘ ´ ˘ ´

And come down heavily like THAT.

´ ˘ ´ ˘ ´ ˘ ´ ˘

Trochees dancing very lightly

´ ˘ ´ ˘ ´ ˘ ´ ˘

Sparkle, froth, and bubble brightly.

´ ˘ ˘ ´ ˘ ˘ ´ ˘ ˘ ´ ˘ ˘

Dactylic daintiness lilting so prettily

´ ˘ ˘ ´ ˘ ˘ ´ ˘ ˘ ´ ˘ ˘

Moves about fluttering rather than wittily.

˘ ˘ ´ ˘ ˘ ´ ˘ ˘ ´ ˘ ˘ ´

While for speed and for haste such a rhythm is the best

˘ ˘ ´ ˘ ˘ ´ ˘ ˘ ´ ˘ ˘ ´

As we find in the race of the quick anapest.

(Boulton 26)

Another metrical foot found in English verse is the *spondee*, where two accented syllables fall together (´ ´), such as in the word "deathbed." Unlike the other metrical feet, however, it's unlikely that the spondee would be used throughout an entire poetic line.

When studying patterns in a metrical line, one other concept that is fundamental is the *number* of feet. Consider, for instance, how many iambs or trochees make up the lines where each term is explained in the above unnamed poem that Marjorie Boulton calls her "four little helps for memory." Since the metrical feet are marked over the lines, it should be easy to count and discover that each line is made up of 4 of the particular feet being described.

When a poetic line is made up of a set number of metrical feet, the following terms are used:

one foot: *monometer*
two feet: *dimeter*
three feet: *trimeter*
four feet: *tetrameter*
five feet: *pentameter*
six feet: *hexameter*
seven feet: *heptameter*
eight feet: *octameter*

Thus, we could say that, lines 1 and 2 are written in iambic tetrameter; lines 3 and 4 are written in trochaic tetrameter; lines 5 and 6 are written in dactylic tetrameter; and lines 7 and 8 are written in anapestic tetrameter.

This quick overview should allow you to be more precise when thinking about the patterns you may have intuitively felt when reading or listening to poetry. We will add to these concepts as we go along.

More about Rhyme

We have already seen how rhyme is created in nursery rhymes when word sounds are repeated at the ends of lines or in predictable patterns within the lines themselves. As with meter, there are many terms that describe rhyme.

Here are a few that you will find helpful:

true rhyme ("perfect" rhyme):	words that rhyme exactly
slant rhyme ("imperfect" rhyme):	words that have close but not identical sounds
masculine rhyme:	rhyme that occurs on stressed syllables
feminine rhyme:	rhyme that occurs on unstressed syllables
eye rhyme:	words that look as though they would rhyme, but in fact are pronounced in such a way that they do not rhyme

internal rhyme: rhyming sounds used in subtle ways within lines of a poem

Edna St. Vincent Millay's poem "Recuerdo" shows how varied and lovely the patterns of rhyme can be. If you read this poem aloud slowly, you will probably first notice how Millay uses pairs of lines that end in true rhyme. Now read the poem again, and listen carefully to the rhyming sounds that you hear within lines or between lines.

We were very tired, we were very merry — a
We had gone back and forth all night on the ferry. a
It was bare and bright, and smelled like a stable — b
But we looked into a fire, we leaned across a table, b
We lay on a hill-top underneath the moon; c
And the whistles kept blowing, and the dawn came soon. c

We were very tired, we were very merry — a
We had gone back and forth all night on the ferry; a
And you ate an apple, and I ate a pear, d
From a dozen of each we had bought somewhere; d
And the sky went wan, and the wind came cold, e
And the sun rose dripping, a bucketful of gold. e

We were very tired, we were very merry, a
We had gone back and forth all night on the ferry. a
We hailed, "Good morrow, mother!" to a shawl-covered head, f
And bought a morning paper, which neither of us read; f
And she wept, "God bless you!" for the apples and the pears, d
And we gave her all our money but our subway fares. d

(Millay *A Few Figs* 10-11)

Although Millay's poem has obvious end-rhyme, as we can hear in the pairs of rhymed lines ("merry" and "ferry," "stable" and "table"), her use of internal rhyme within lines 1 through 4, created by the words "tired," "night," "bright," and "fire," adds a nostalgic, song-like quality to this reminiscence. Since these internal rhymes also fall on stressed syllables, there is an extra moment of attention paid to the details of the scene. In fact, the internal rhymed lines create a second level to the narrative or story being told in the poem, where the words "tired . . . night . . . bright . . . fire" emphasize the intimacy and physical sensations of the scene. In the remaining stanzas of the poem, Millay's use of sound becomes even more complex.

In the second stanza, she uses a combination of *assonance* (the repetition of a vowel sound) and rhyme in line 9, with the recurring use of short and long "a" sounds: *And you ate an apple, and I ate a pear.* And in the third stanza, the imperfect rhymes of "morrow," "mother," and "morning" serve to draw lines 15 and 16 together through both sound and image.

The conventional method for indicating the rhyme pattern in a series of lines is to label the sound at the end of the first line as "a" and identify each similar subsequent end-line sound as "a." Each new sound is labeled with successive letters of the alphabet ("b," "c," "d," and so on), as shown in the margin of "Recuerdo."

Traditional Forms: Limerick and Ballad

You can probably imagine the countless ways in which patterns of sound and meter could be combined in poetry. From the many possible patterns, certain specific combinations have evolved and endured across cultures and over time to become what we know as traditional poetic forms. The word *form* here refers to a type of poetry which has certain set characteristics in its use of rhyme and/or meter, and sometimes even in its subject matter. Some traditional forms of English poetry are studied but not composed very often today; others, such as the sonnet, have enjoyed long and continued popularity. Some forms have also been the subject of great experimentation and variation, as we will see in Chapter 2. Let's begin with some traditional forms well known beyond the poetry classroom.

Perhaps closest to the nursery rhyme in its humor and rollicking meter is the limerick. Here's an example from Edward Lear who helped popularize the form:

There was an Old Man of the Coast,
Who placidly sat on a post;
But when it was cold he relinquished his hold,
And called for some hot buttered toast.

(Lear 93)

In the limerick, the first two lines rhyme with the last line, all of which have three stressed beats. In this example from Lear's *The Complete Nonsense Book*, line 3 features internal rhyme ("cold" and "hold") and four stressed beats. Often limericks are written in five lines, breaking the third line into two separate lines. In this case, lines 3 and 4 rhyme and have two stressed beats. Regardless of whether it is printed in 4 or 5 lines on the page, the easily-recognized meter of a limerick is almost certain to invite a laugh even when chanted in nonsense syllables. The subject matter of limericks is usually humorous, sometimes irreverent or bawdy. The limerick has even been used to make light of mathematics:

> There was a young man from Trinity
> Who solved the square root of infinity.
> While counting the digits,
> He was seized by the fidgets,
> Dropped science, and took up divinity.
>
> (Fadiman 295)

Another traditional poetic form, the ballad, satisfies our ear with familiar patterns, while often telling a tale of lost love, battle, or betrayal. The traditional Scottish or English ballad is written in four-line stanzas. Lines 1 and 3 have four stressed syllables; lines 2 and 4 are rhymed and have three stressed syllables. "Sir Patrick Spens" is a well-known anonymous Scottish ballad of treachery and shipwreck; it is also sometimes titled "Sir Patrick Spence."

> The king sits in Dumferling toune,
> Drinking the blude-reid wine:
> "Oh whar will I get guid sailor,
> To sail this schip of mine?"
>
> Up and spak an eldern knicht,
> Sat at the kings richt kne:
> "Sir Patrick Spence is the best sailor
> That sails upon the se."
>
> The king has written a braid letter,
> And signd it wi his hand,
> And sent it to Sir Patrick Spence,
> Was walking on the sand.
>
> The first line that Sir Patrick red,
> A loud lauch lauched he;
> The next line that Sir Patrick red,
> The teir blinded his ee.
>
> "O wha is this has done this deid,
> This ill deid don to me,
> To send me out this time o' the yeir,
> To sail upon the se!

"Mak hast, mak haste, my mirry men all,
Our guid schip sails the morne:"
"O say na sae, my master deir,
For I feir a deadlie storme.

"Late late yestreen I saw the new moone,
Wi the auld moone in hir arme,
And I feir, I feir, my deir master,
That we will cum to harme."

O our Scots nobles wer richt laith
To weet their cork-heild schoone;
Bot lang owre a' the play wer playd,
Their hats they swam aboone.

O lang, lang may their ladies sit,
Wi thair fans into their hand,
Or eir they se Sir Patrick Spence
Cum sailing to the land.

O lang, lang may the ladies stand,
Wi thair gold kems in their hair,
Waiting for thair ain deir lords,
For they'll se thame na mair.

Haf owre, haf owre to Aberdour,
It's fiftie fadom deip,
And thair lies guid Sir Patrick Spence.
Wi the Scots lords at his feit.

("Sir Patrick Spens" 65-67)

The rhyming pattern is easily seen here, as are some of the other features common to the ballad. For instance, the anonymous storyteller uses *incremental change*, the slight changing of detail in a repeated phrase, ("lang, lang may their ladies sit" and "lang, lang may their ladies stand" in stanzas 9 and 10) to add to the tension of the widows' waiting and to indicate the passage of time. As with many ballads, the critical scene in the story — the actual shipwreck — is not described directly; instead, we hear about the circumstances that lead up to the storm and see the aftermath both under the sea and on shore.

This poem, too, has been the subject of a mathematical spoof in Arthur T. Quiller-Couch's thirty-four stanza invention "A New Ballad of Sir Patrick

Spens." As you can see, the first stanza sets up a parody of the traditional ballad. Rather than looking for a good sailor, the king is now seeking someone with talent in geometry in this extended "in-joke" for the mathematically savvy.

The King sits in Dunfermline town
  Drinking the blude-red wine:
"O wha will rear me an equilateral triangle
  Upon a given straight line?"

Of course, it is Sir Patrick Spens who will eventually prove

. . .That things was equal to the same.
  Was equal ane to t'ither. . .

His proof, in part, reads as follows:

"Sith in the circle first I drew
  The lines BA, BC,
By radii true, I wit to you
  The baith maun equal be."

(Quiller-Couch 261-65)

Later, we will have more to say about the tradition of commemorating mathematical achievements in verse, of which this ballad is a clever and slightly irreverent example.

Patterns in Ideas: Counting and Listing

In addition to patterns that may be felt or heard in a poem, there are other patterns that may be "counted." Such patterns may be found in the arrangement of ideas. Whether in a traditional poetic form, or in lines whose elements of meter or rhyme are less formal, poets often use subtle patterns in language to itemize, build, and unify the poem's ideas. This strategy is very effective and is found in dozens of familiar poems. For example, after Elizabeth Barrett Browning asks, in her most famous sonnet, "How do I love thee?," she goes on to "count the ways." Browning uses vivid metaphorical language to measure her love alongside the most *immeasurable* things she can imagine. To do this, she ticks off familiar actions (reaching as far as one can in all directions) and examples of idealized behavior (striving for Right, turning from Praise), as she tries to quantify that which is boundless.

How do I love thee? Let me count the ways.
I love thee to the depth and breadth and height
My soul can reach, when feeling out of sight
For the ends of Being and ideal Grace.
I love thee to the level of everyday's
Most quiet need, by sun and candle light.
I love thee freely, as men strive for Right;
I love thee purely, as they turn from Praise.
I love thee with the passion put to use
In my old griefs, and with my childhood's faith.
I love thee with a love I seemed to lose
With my lost saints, — I love thee with the breath,
Smiles, tears, of all my life! — and, if God choose,
I shall but love thee better after death.

(Browning 43)

In this case, the patterns created by the poem's ideas are easy to follow. They build in passion as the poem unfolds, and gain intensity with each of the nine repetitions of "I love thee."

In his poem "The Negro Speaks of Rivers," Langston Hughes is also using listing patterns in the language of the poem. This poem draws on images beyond Hughes' own experience to forge a link with people who have come before him. The patterns of repetition occur within the poem's simple parallel structure of repeated subject-verb phrases. Rather than conforming to a formal pattern of rhyme or using the same meter or line length throughout the poem, Hughes relies on a different kind of structure to give the poem its unity. Through the listing of actions, ("I've known... I bathed... I built... I looked... I looked... I heard... I've known..."), the lines flow further and further out from the beginning, reminding us of a mathematical sequence, as the timelessness of rivers connects the experiences of many people and civilizations. Like currents in a river, continually moving but contained within two banks, the poem's lines change continually while maintaining a subtle, unifying pattern. Hughes' poem is an excellent example of a modern poem written in free verse whose structure comes not from a regularized metrical pattern, but from a more fluid, deeper sense of structure that casts off easily-measured ideas of meter.

I've known rivers:
I've known rivers ancient as the world and older than the
flow of human blood in human veins.

My soul has grown deep like the rivers.

I bathed in the Euphrates when dawns were young.
I built my hut near the Congo and it lulled me to sleep.
I looked upon the Nile and raised the pyramids above it.
I heard the singing of the Mississippi when Abe Lincoln
    went down to New Orleans, and I've seen its muddy
    bosom turn all golden in the sunset.

I've known rivers:
Ancient, dusky rivers.

My soul has grown deep like the rivers.

(Hughes 23)

The essential patterns in rhythm, rhyme, and ideas we have seen in these poems are in many ways comparable to the countable patterns of mathematics. Let's continue by looking closely at more of the forms created by the patterns in mathematics and poetry.

Chapter 2 – Counting Patterns Take Form

2.1 How Patterns Take Form in Mathematics

Three of the most famous patterns associated with numbers are Pascal's Triangle, Fibonacci Numbers, and the Golden Ratio. Describing one of them inevitably leads to mention of the other two, since they are inextricably linked together in both simple and theoretical ways. We will examine how mathematicians uncover such patterns, explore how they play with patterns to find new relationships, and search for instances of these patterns outside the domain of mathematics. This section provides just a taste of these three celebrated patterns. We hope it will be a tantalizing one that will lead the reader to explore them more deeply.

Pascal's Triangle

Pascal's Triangle is an ancient number pattern that is still very important to modern mathematical theory. Although it is easy to construct the triangle, there are many complex patterns within its simplicity. A. W. F. Edwards begins his book *Pascal's Arithmetic Triangle* by calling it "the most famous of all number patterns" (xii).

A very early reference to this arrangement of numbers exists in a tenth-century work by the mathematician al-Karaji of Baghdad. Others may be found in the eleventh-century work of the Persian mathematician-poet Omar Khayyám and in the thirteenth-century work of the Chinese mathematician Yang Hui. But it was Blaise Pascal's careful analysis in his 1654 book *Traite du Triangle Arithmetique* that resulted in his name being associated with the triangle. Pascal explored both the pure mathematics found in the triangle, as well as its application to probability theory. Although these numbers have been arrayed in a variety of forms throughout the ages, some forms more rectangular than triangular, the layout of the first few rows of what we now refer to as "Pascal's Triangle" is shown below. Note that the top row, by common convention, is numbered "Row 0."

Row 0	1
Row 1	1 1
Row 2	1 2 1
Row 3	1 3 3 1
Row 4	1 4 6 4 1
Row 5	1 5 10 10 5 1
Row 6	1 6 15 20 15 6 1
Row 7	1 7 21 35 35 21 7 1
⋮	⋮

Figure 2.1

How does this pattern work and what aesthetic and mathematical attributes arise from it? The number 1 forms the outside edges of each row. Each interior number in a row is the sum of the two nearest numbers in the row above, as illustrated with Rows 4 and 5 in Figure 2.2.

Row 4	1 4 6 4 1
Row 5	1 5 10 10 5 1

Figure 2.2

We see in the fifth row that the two instances of the number 5 are the sum of 1 and 4 in the row above, while the two instances of the number 10 are the sum of 6 and 4 in Row 4. Another important pattern is found in the relationship between the numbers in Pascal's Triangle and the coefficients of the expansion of $(a+b)^n$, as illustrated in Figure 2.3. Note that the coefficients, typed in bold in the expansion column, and the numbers in the triangle are identical.

Binomial	Expansion	Pascal's Triangle
$(a+b)^0 =$	1	1
$(a+b)^1 =$	$1a + 1b$	1 1
$(a+b)^2 =$	$1a^2 + 2ab + 1b^2$	1 2 1
$(a+b)^3 =$	$1a^3 + 3a^2b + 3ab^2 + 1b^3$	1 3 3 1
$(a+b)^4 =$	$1a^4 + 4a^3b + 6a^2b^2 + 4ab^3 + 1b^4$	1 4 6 4 1
$(a+b)^5 =$	$1a^5 + 5a^4b + 10a^3b^2 + 10a^2b^3 + 5ab^4 + 1b^5$	1 5 10 10 5 1

Figure 2.3

Also, observe the pattern in any arbitrary row, for instance Row 5, where we are looking at the expansion of $(a+b)^5$. As we move from left to right through the terms of the expansion, the powers on the variable *a* descend from 5 to zero, while the powers on *b* rise from zero to 5 for each term of this row. Further, the sum of the powers on *a* and *b* in any term of row 5 must add up to 5. This type of pattern is true for every row. We can now predict the pattern of terms and their coefficients for the expansion of a binomial such as $(a+b)^{10}$ by continuing Pascal's triangle to the 10th row. The 10th row of Pascal's Triangle is

1 10 45 120 210 252 210 120 45 10 1

so we can predict that $(a+b)^{10} =$

$$1a^{10} + 10a^9b + 45a^8b^2 + 120a^7b^3 + 210a^6b^4 + 252a^5b^5 + 210a^4b^6 + 120a^3b^7 + 45a^2b^8 + ab^9 + 1b^{10}$$

But there is much more to the pattern of this triangle. The sum of the numbers in row zero is 1 which equals 2^0; the sum of the numbers in row one is 2 or 2^1; the sum of the number in row two is 4 or 2^2. This pattern continues, so we say the sum of the numbers in row *n* is 2^n. For example, if we add up the numbers in the 7th row of Pascal's Triangle, the sum should be 2^7 or 128. Let's work our way down the triangle to the 7th row and see if the pattern holds.

Row 0	1	1	$= 2^0$
Row 1	1 1	1+1 = 2	$= 2^1$
Row 2	1 2 1	1+2 +1 = 4	$= 2^2$
Row 3	1 3 3 1	1+3+3+1 = 8	$= 2^3$
Row 4	1 4 6 4 1	1+4+6+4+1 = 16	$= 2^4$
Row 5	1 5 10 10 5 1	1+5+10+10+5 +1 = 32	$= 2^5$
Row 6	1 6 15 20 15 6 1	1+6+15+20+15+6+1 = 64	$= 2^6$
Row 7	1 7 21 35 35 21 7 1	1+7+21+35+35+21+7+1=128	$= 2^7$

Figure 2.4

There are many other patterns exhibited here, as well. First, let's add up the numbers along diagonal lines, as in Figure 2.5 below. The sums are given on the right side of the triangle.

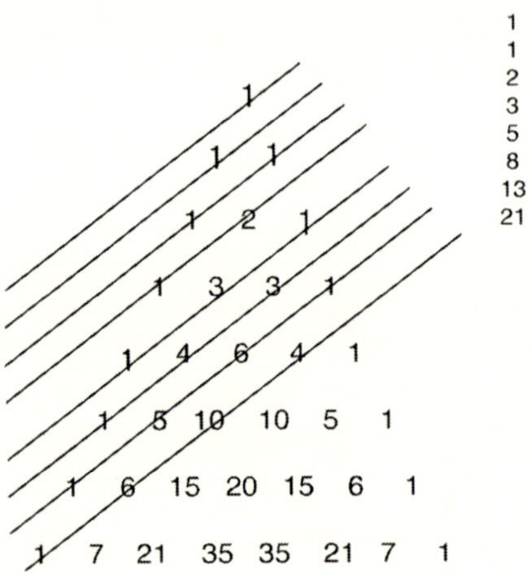

Figure 2.5

The following sequence of numbers represents the sums of these diagonals:

1, 1, 2, 3, 5, 8, 13, 21, . . .

What is so special about this sequence? Notice that after the first two 1's, each number is the sum of the two previous numbers (2 = 1 + 1; 3 = 1 + 2; 5 = 2 + 3; etc.) This sequence is the Fibonacci Sequence or the Fibonacci Numbers, named after the twelfth-century Italian mathematician, Leonardo of Pisa, known colloquially as Fibonacci or "son of Bonacci." We will discuss the Fibonacci Numbers and their close cousins, the Lucas Numbers, later in this section.

Let's return to Pascal's Triangle to search for additional patterns. The Triangle holds many number patterns, including the listing of the Natural Numbers, the powers of the number 11, and the different combinations formed by choosing subsets of *n* objects. Entire books are devoted to exploring patterns within Pascal's Triangle. To get a flavor for the variety and scope of these patterns, peruse the charts in Figure 2.6 where some of the triangle's best-known patterns are illustrated. Some are easy to find and understand, while others may tax your mathematical understanding.

The Natural Numbers	1, 2, 3, 4, 5, 6, 7, . . . The Naturals Numbers are found along the boxed diagonals.	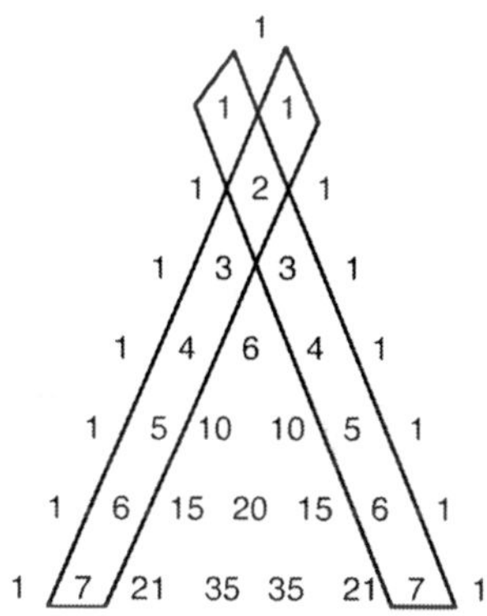

Powers of 11	1, 11, 121, 1331,14641,. . . From row 0 to row 4, read across each row to find the *n*th power of 11 in the *n*th row. After row 4, interpret the digits in the given row to be placeholders of powers of 10. For instance in the 5th row: 1 5 10 10 5 1 = $1 \times 10^5 = 100000$ $+ 5 \times 10^4 = 50000$ $+ 10 \times 10^3 = 10000$ $+ 10 \times 10^2 = 1000$ $+ 5 \times 10^1 = 50$ $+ 1 \times 10^0 = 1$ $= 161051 = 11^5$	1 $= 11^0$ 1 1 $= 11^1$ 1 2 1 $= 11^2$ 1 3 3 1 $= 11^3$ 1 4 6 4 1 $= 11^4$ 1 5 10 10 5 1 1 6 15 20 15 6 1 1 7 21 35 35 21 7 1
Combinations	To find the number of ways to choose k objects from a total of n objects, find the *kth* entry in the *nth* row of the Triangle. For example: There are 10 ways to choose 3 objects from a total of 5 objects – this is also known as the number of combinations of 3 objects chosen from set of 5. The number 10 is found by going to the 3rd entry of the 5th row.	1 Row 0 1 1 Row 1 1 2 1 Row 2 1 3 3 1 Row 3 1 4 6 4 1 Row 4 1 5 (10) 10 5 1 Row 5 1 6 15 20 15 6 1 Row 6 1 7 21 35 35 21 7 1 Row 7

Triangular Numbers	Triangular Numbers: 1, 3, 6, 10, 15, . . . The *nth* Triangular number is the number of dots in a triangle whose base has as many dots as the *nth* Natural Number. For example, when $n = 3$, three dots make up the base of a triangle whose total number of dots is 6. The third triangular number is 6. 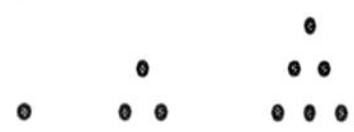 Dots in Base 1 2 3 Dots in Triangle 1 3 6	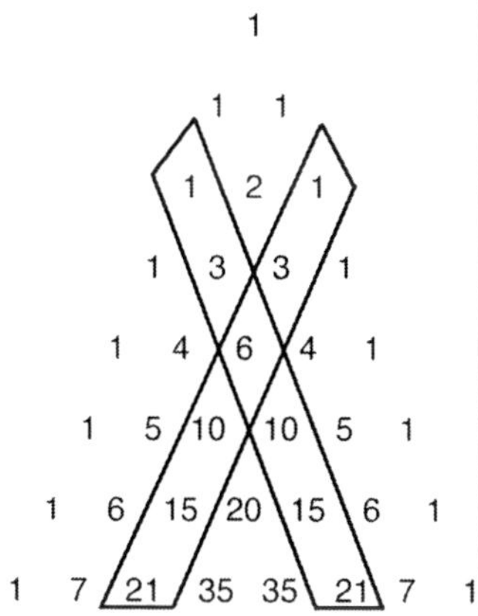
Catalan Numbers	Catalan Numbers: 1, 2, 5, 14, 42, 132, 429, . . . The *n*th Catalan Number represents the number of ways to divide a polygon with *n+2* sides into *n* triangles using non-intersecting diagonals. To find the *n*th Catalan Number in Pascal's Triangle, take the *n*th circled number on the center column and subtract the boxed number to its right. The 4th Catalan Number (14) is found by taking the 4th number on the center column (70) and subtracting the number to its right (56).	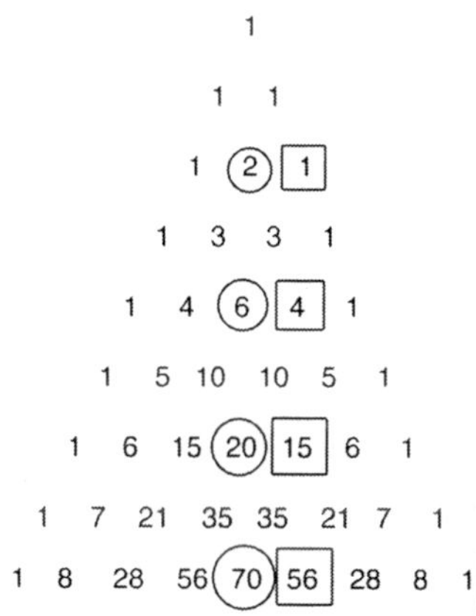

Figure 2.6

Figure 2.7 (below) demonstrates the relationship between each Catalan Number, its position in the sequence, and its corresponding polygon.

Sequence	a_1	a_2	a_3	a_4	a_5	a_6	a_7	...
Position in Sequence (n)	1	2	3	4	5	6	7	...
Number of Sides on Polygon ($n+2$)	3	4	5	6	7	8	9	...
Catalan Number - the number of ways to divide a polygon with $n+2$ sides into n triangles using non-intersecting diagonals	1	2	5	14	42	132	429	...

Figure 2.7

The Fibonacci Sequence

One of the more intriguing patterns found in Pascal's Triangle is the Fibonacci Sequence. Remember that after the first two terms (both 1's), each subsequent term in the sequence is the sum of the two previous terms. This is called a recursive sequence, since each new entry is defined in terms of the previous entries. We can write down a recursive definition for each Fibonacci number F_n.

$$F_1 = F_2 = 1$$
$$F_{n+1} = F_n + F_{n-1} \text{ for } n \geq 2$$

A table of the first 20 Fibonacci Numbers is displayed below in Figure 2.8.

F_1	1	F_6	8	F_{11}	89	F_{16}	987
F_2	1	F_7	13	F_{12}	144	F_{17}	1597
F_3	2	F_8	21	F_{13}	233	F_{18}	2584
F_4	3	F_9	34	F_{14}	377	F_{19}	4181
F_5	5	F_{10}	55	F_{15}	610	F_{20}	6765

Figure 2.8

A second well-known sequence of integers, the Lucas Numbers, named for the French mathematician Edouard Lucas, is very similar to the Fibonacci Numbers, but begins with the numbers 1 and 3:

1, 3, 4, 7, 11, 18, 29, 47, . . .

We can write also write a recursive definition for the Lucas Numbers.

$$L_1 = 1$$
$$L_2 = 3$$
$$L_{n+1} = L_n + L_{n-1} \text{ for } n \geq 2$$

Other integer sequences can be similarly constructed, but for the purposes of this book, we will confine our discussion to the Fibonacci Sequence because it will lead us to the third mathematical pattern we will discuss, namely the Golden Ratio.

About 800 years ago, Leonardo of Pisa, or Fibonacci, first introduced this sequence in his text *Liber Abaci*. The text was influential in moving European mathematics from the cumbersome system of Roman numerals (I, II, III, IV, . . .) to Arabic numerals (1, 2, 3, 4, . . .). It also contained a fanciful problem about breeding rabbits that should remind us of the riddle-poem in Chapter 1 about the man going to St. Ives. The problem asked how many rabbits would be produced in one year from a single pair of rabbits, if the initial pair produces a new pair every month and each new pair begins similar breeding in their second month after birth. In this make-believe world, none of the rabbits dies. The sequence, known today by Fibonacci's name, is the list of the number of pairs of rabbits that exist each month.

Let's see how this breeding pattern works and how quickly the rabbits multiply, as shown in Figure 2.9. In just twelve months' time there are 144 breeding pairs (the 12th Fibonacci Number); at the end of two years there are 46,368 breeding pairs (the 24th Fibonacci Number); and in 5 short years there would be over 1.5 trillion breeding pairs (the 60th Fibonacci number is 1,548,008,755,920). And so we see the pattern emerge:

1, 1, 2, 3, 5, 8, 13, 21, 34, 55, 89, . . .

Time	Explanation	Total Number of Pairs
Initially	1 pair	1 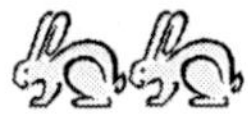
January	1 pair	1 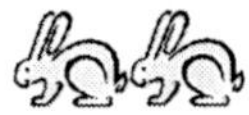
February	The original pair of rabbits produces a new pair.	2 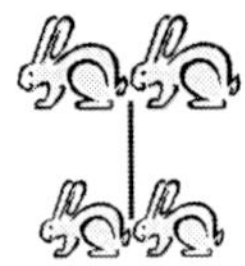
March	The original pair of rabbits produces their second new pair.	3 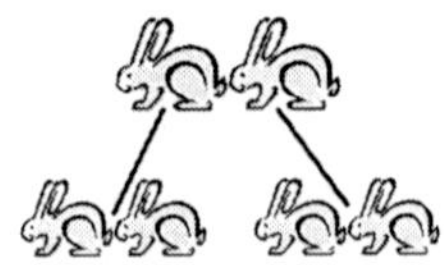
April	The original pair of rabbits produces yet another pair, while the pair of rabbits born in February produces its first offspring.	5 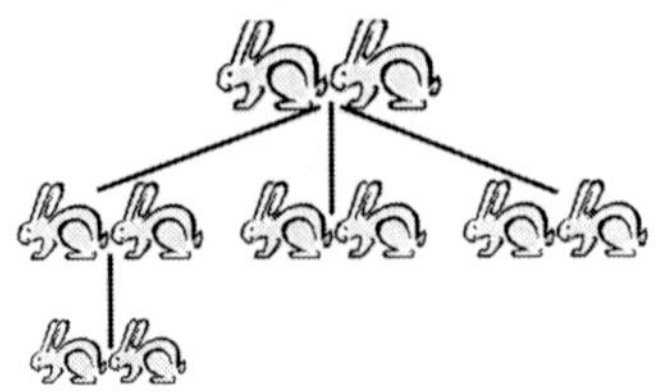
May	The original pair is still producing one new pair each month – this will be their 4th set. The pair of rabbits born in February produces its 2nd set and the pair born in March gives birth to its 1st set.	8 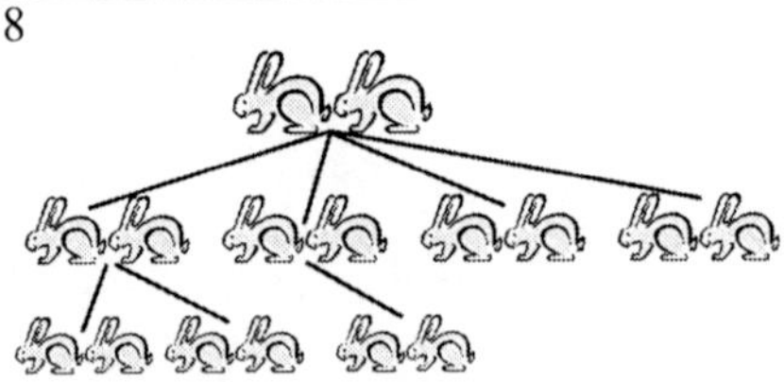

Figure 2.9

Fibonacci numbers are intriguing, not only because of their pattern, but also because this particular sequence of numbers occurs in a variety of natural settings. For some wildflowers, the number of petals (or petal-like parts) is a Fibonacci number. Flowers in the Birthwort, Iris, Lily, Spiderwort, and Water Plantain families have 3 petals or 3 lobes. Many of these flowers also have 3 sepals that often look like petals. A remarkable number of wildflowers have 5 petals, 5 petal-like sepals, or 5 lobed flowers. These include members of the Bladderwort, Borage, Buttercup, Flax, Geranium, Mallow, Phlox, Pink, Primrose, Rose, and Saxifrage families.[3] The wildflowers in Figure 2.10 all exhibit patterns of Fibonacci numbers.

Mountain Iris (Iris Family) 3 petals, 3 down-curving sepals Yellowstone Nat. Park, WY	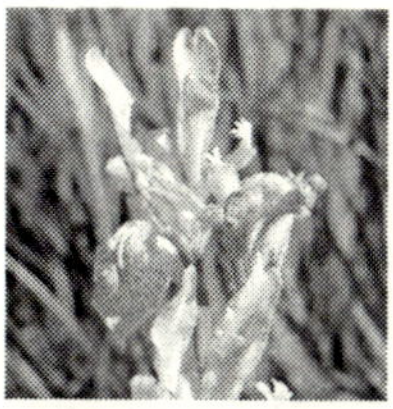	Painted Trillium (Lily Family) 3 petals, 3 sepals Adirondack State Park, NY	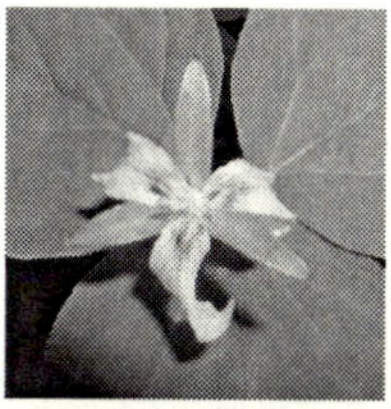
Sego Lily (Lily Family) 3 pointed petals Grand Teton Nat. Park, WY	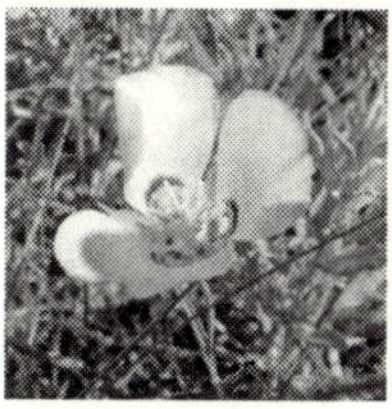	Alpine Forget-me-not (Borage Family) 5 lobed flower Yellowstone Nat. Park, WY	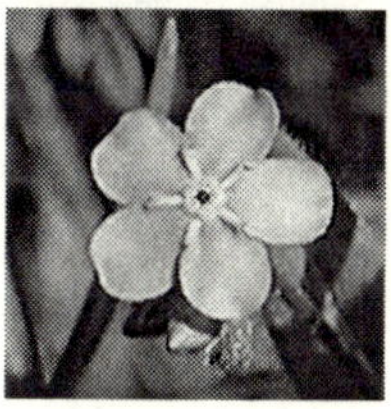
China Flower (Citrus Family) 5 petals with 5 red suffusions Western Cape Peninsula, South Africa	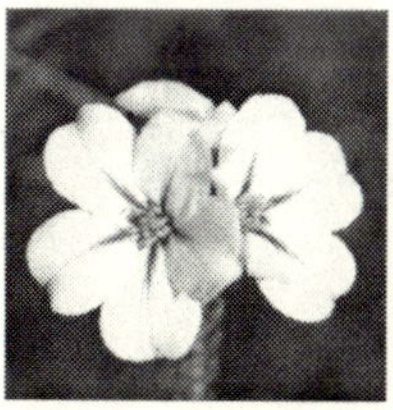	Butterwort (Bladderwort Family) 5 petal-like lobes of unequal length Thingvellir Nat. Park, Iceland	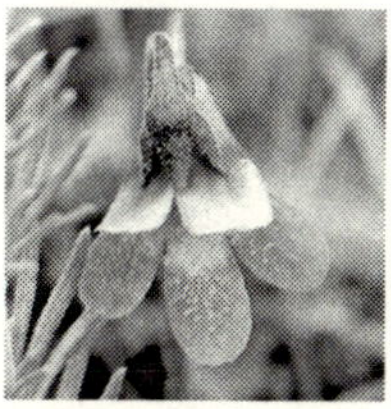
Yellow Columbine (Buttercup Fam.) 5 tubular petals, 5 petal-like sepals Grand Teton Nat. Park, WY	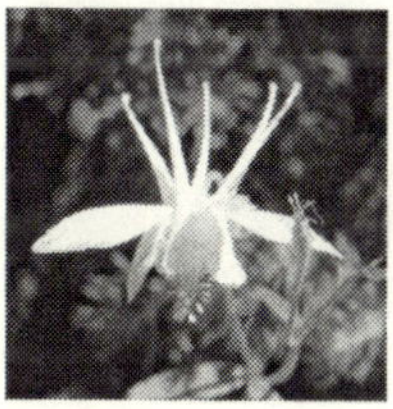	Wood Cranesbill (Geranium Family) 5 petals, 5 sepals Thingvellir Nat. Park, Iceland	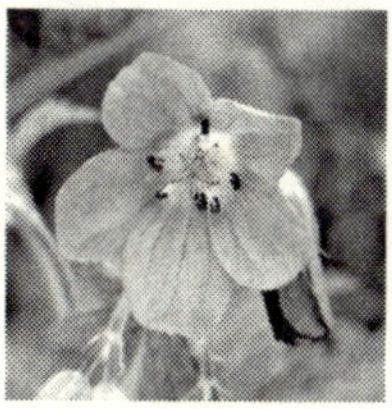

Figure 2.10

[3] Steven Daniel, Nature Discoveries, Inc., Rochester, NY, assisted with wildflower taxonomy.

Fibonacci Numbers are also found in the Aster family, a grouping that includes asters, black-eyed Susans, daisies, coneflowers, and sunflowers. Some members of the Aster family are called "composites," since they actually have two arrangements of flowers: a set of ray flowers that look like petals and a set of tiny, central disk flowers surrounded by the rays. In some composites, the number of ray flowers tends to be a Fibonacci Number, such as 8, 13, 21, 54 or 89, but there is much natural variation. Examples of composites that often have 8 ray flowers include Lance-leaved Coreopsis, Nodding Bur Marigold and Tickseed Sunflower.

We should not overemphasize the relationship between the number of petals or rays and Fibonacci Numbers, since there are many wildflowers with 4 or 6 petals (NOT Fibonacci numbers) and many composites whose rays appear in a wide range of values.

But there is a second, more important Fibonacci relationship in the spirals formed by the disk flowers of some composite flowers, a relationship that is echoed in spirals found in pinecones, pineapples, and artichokes. Composites such as daisies, coneflowers, and sunflowers have central disks arranged so that two or more sets of spirals are formed by their disk florets. The numbers of clockwise and counterclockwise spirals are almost always consecutive Fibonacci Numbers. The consistency of formation and number of spirals is a result of efficient seed packing in the growth of the central disk. Look carefully at the daisy in Figure 2.11 to see if you can discover two sets of spirals. These spirals are traced out in Figure 2.12, where 21 spirals are shown curving in a clockwise direction on the left, and 34 spirals are shown curving in a counterclockwise direction on the right.

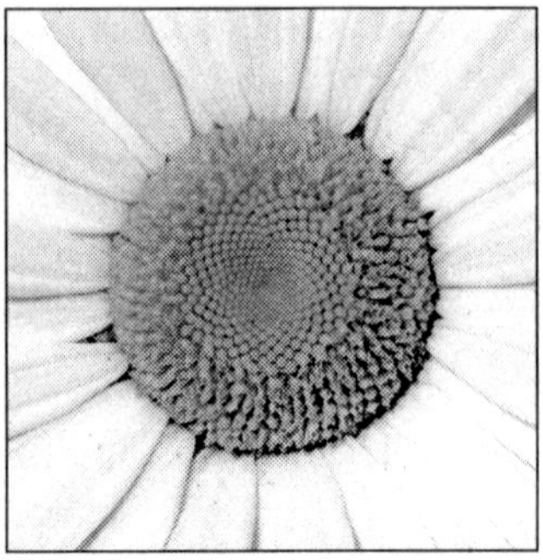

Figure 2.11

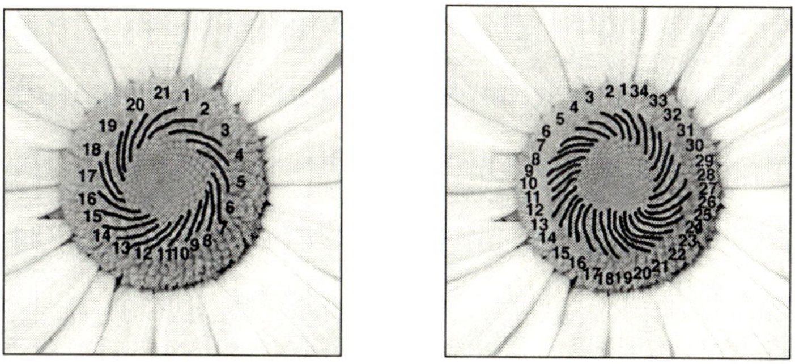

Figure 2.12

Like the disks of daisies, the bracts at the base of pinecones display two sets of spirals, one going clockwise and the other counterclockwise. As in composites, the numbers of spirals in each set are usually consecutive Fibonacci numbers. In Figure 2.13 we see the base of a pinecone that has 8 spirals traced out in one direction and 13 spirals in the other.

Pinecone Base | 8 Clockwise Spirals | 13 Counterclockwise Spirals

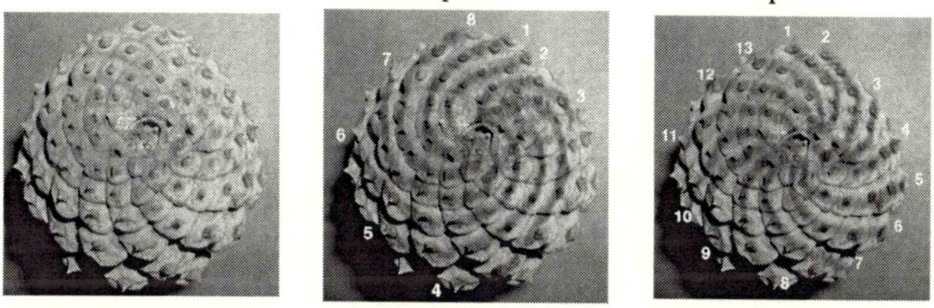

Figure 2.13

Pineapples, too, exhibit this strange property in their spirals: the number of spirals in each direction occur in consecutive Fibonacci Numbers. Since the scales of a pineapple are roughly hexagonal in shape, they produce three different types of spirals, each type passing through two opposite sides of the "hexagon." The leftmost photograph in Figure 2.14 shows one of the steepest spirals. On this pineapple there are 21 parallel steep spirals. The middle photo of the same pineapple shows a second type of spiral rising moderately from right to left. There are 13 parallel moderate spirals. In the photo on the right, a third type of gradual spiral is shown rising from left to right. There are 8 gradual parallel spirals on this pineapple.

Other pineapples will similarly display three sets of spirals that appear in 13 steep, 8 moderate, and 5 gradual sets. Once again, the tendency in all pineapples is to exhibit three sets of spirals in consecutive Fibonacci numbers.

21 Steep Spirals   13 Moderate Spirals   8 Gradual Spirals

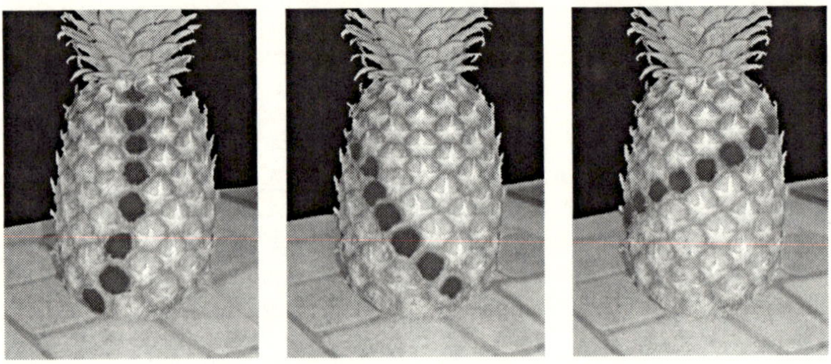

Figure 2.14

A final illustration shows two sets of spirals, one steep and one gradual, appearing in the leaves of an artichoke. On the left of Figure 2.15 we see one highlighted example of the 8 parallel rows of artichoke leaves that spiral up steeply. On the right, the photograph shows one highlighted example of the 5 parallel rows of leaves that rise more gradually from left to right.

8 Steep Spirals   5 Gradual Spirals

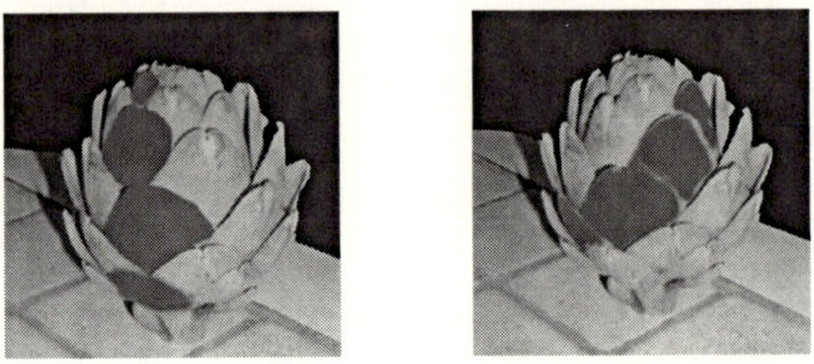

Figure 2.15

As with the pinecone and pineapple, the spirals in artichoke leaves almost always occur in sets of consecutive Fibonacci numbers.

Let's move to an entirely different area where Fibonacci Numbers may appear — the sphere of music. An octave consists of 8 notes and is represented on the piano by 8 keys, illustrated in Figure 2.16 by a C-octave.

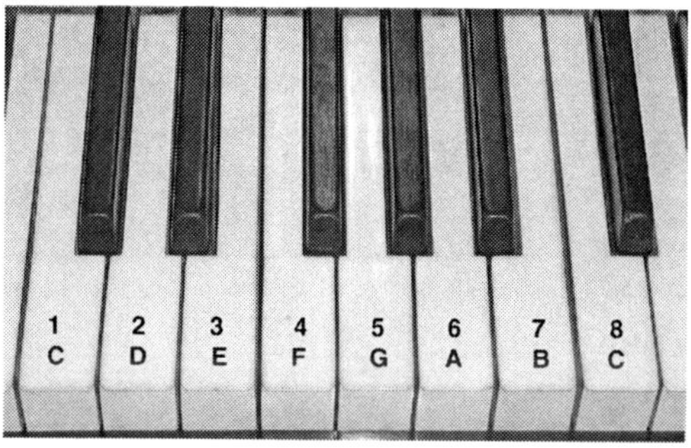

Figure 2.16

If we include sharps and flats, we add 5 black keys to the 8 white keys for a total of 13 keys, often referred to as the chromatic scale. The black keys themselves are positioned in groups of 2 and 3. All the numbers mentioned — 2, 3, 5, 8, and 13 — are Fibonacci Numbers. But as with wildflowers, we must be careful about selectively examining some patterns of the keyboard while ignoring others. For example, although the chromatic scale is usually pictured as shown above in Figure 2.16, the chromatic scale, in fact, consists of only 12 pitches, not 13. The twelve pitches are represented by the twelve white and black keys from C (on the left) up to and including B (on the right), each a half tone apart.

Some authors have found Fibonacci-like patterns or Fibonacci ratios in the works of composers as diverse as Mozart, Beethoven, Bach, Schubert, Debussy, and Bartók, while others challenge these claims. Mario Livio presents an excellent summary of the research in this area in his book *The Golden Ratio.* Although he discounts many of the claims, he supports the idea that some twentieth-century composers have shown a strong interest in using numbers in their compositions, including Fibonacci ratios (Livio 183-94).

In poetry, too, occurrences of Fibonacci Numbers are linked to familiar patterns. The common limerick, first discussed in Chapter 1, is composed of 5 lines with 13 stressed beats. These beats are found in patterns of 2 and 3:

Line	Stress on Capitalized Words	Stressed Beats
1	There WAS a young MAN from TRINity	3
2	Who SOLVED the square ROOT of inFINity	3
3	While COUNTing the DIGits	2
4	He was SEIZED by the FIDgets	2
5	Dropped SCIENCE, and TOOK up diVINity.	3
5 Lines		13 Beats

(Fadiman 295)

Turning from the limerick to more serious poetry, scholars have studied the appearance of the Fibonacci Sequence, or ratios of successive Fibonacci Numbers, in epic works including Virgil's *Aeneid*, written in the first century B.C.E., and in Dante's *The Divine Comedy*, written in the thirteenth century C.E. George E. Duckworth conducted a well-known study published in 1962 titled *Structural Patterns and Proportions in Vergil's Aeneid: A Study in Mathematical Composition*. Duckworth examines the structure of the entire *Aeneid*, as well as the patterns of its individual books, and finds hundreds of instances of Fibonacci ratios, or what we refer to in the next section of this chapter as the *Golden Mean*. He feels that these numerous instances of the Golden Mean illuminate Virgil's[4] method of composition and resolve issues of interpretation of specific passages. While some contemporary authors, such as Thomas Koshy, support Duckworth's ideas, others agree with Roger Herz-Fischler's research that disputes Duckworth's mathematical methods.

While we may not know for certain what Virgil intended, we do know that some contemporary poets are experimenting with the Fibonacci Sequence in the number of words used in each line, in the number of syllables per line, or in new fixed forms that incorporate parts of the Fibonacci sequence. Some examples include "Fibonacci Memory," written by Robert M. Wilson, where the number of lines in each of five stanzas is a consecutive Fibonacci Number; the prose-poem *Tjanting*, by Ron Silliman, where the number of lines in each of nineteen stanzas or paragraphs is a consecutive Fibonacci Number; and the book-length poem *Alphabet*, by the Danish poet Inger Christensen, that combines the alphabet and Fibonacci Numbers in its structure. Michael L. Johnson's poem "Fibonacci Time Lines" is an example of syllabic verse where the syllable count for each line is a consecutive Fibonacci Number. Interestingly, each item named in this poem has a relationship to Fibonacci Numbers.

[4] Although Duckworth used the Latin spelling of "Vergil" in his title, we have chosen to use the contemporary English spelling of "Virgil."

cat's
claw's
curl, pine-
cone's swirl, goat's
horn's turn, nautilus'
shell's homing out, pineapple's whorl,
sneezewort's branchings, hair's twist, parrot's beak's
    growth, elephant's
tusk's curve, monkey's tail's spiral, cochlea's whirl
    of sound, Vitruvius' analogies,
Parthenon's geometry, logarithms' golden sections, time's
    way through form, mind's acceleration on its helical
    vector to death . . .

(Johnson "Fibonacci" 80)

The Golden Ratio or Golden Number, Φ

Just as Pascal's Triangle led us to the patterns of Fibonacci Numbers, the Fibonacci Numbers now lead us to the last number pattern we will discuss, the Golden Ratio. This ratio, known by the Greek letter "phi" or Φ, is an irrational number whose exact value is $(1+\sqrt{5})/2$, and is approximated by 1.61803398875. . . Euclid defined the Golden Ratio around 300 BCE. He began by drawing a line that is divided into two segments. The ratio of the larger segment to the smaller is the same value as the ratio of the entire line to the larger segment. The diagram in Figure 2.17 illustrates this ratio.

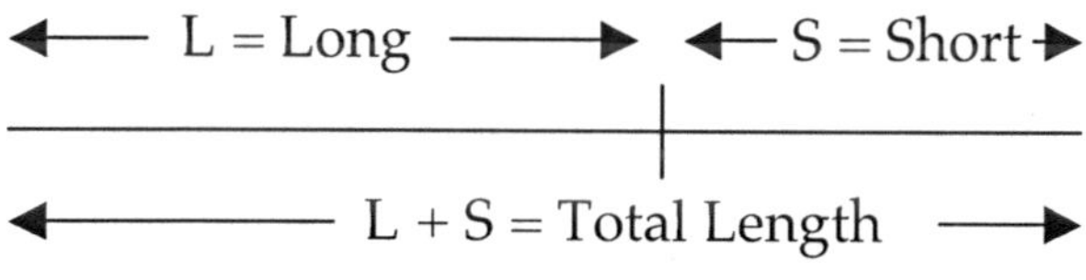

$$\Phi = \frac{L}{S} = \frac{L+S}{L} = 1.61803398875...$$

Figure 2.17

In addition to the Golden Ratio, Φ has many other names, including the Golden Number, the Golden Section, the Golden Mean, and the Divine

Proportion. How are Fibonacci Numbers and the Golden Ratio linked? Examine the table of Fibonacci Ratios (F_{n+1}/F_n) shown in Figure 2.18. These ratios are formed by dividing successive Fibonacci Numbers and rounding the answer to 7 decimal places.

$\frac{F_2}{F_1}$	$\frac{1}{1}=1$	$\frac{F_7}{F_6}$	$\frac{13}{8}=1.625$	$\frac{F_{12}}{F_{11}}$	$\frac{144}{89}=1.6179775$
$\frac{F_3}{F_2}$	$\frac{2}{1}=2$	$\frac{F_8}{F_7}$	$\frac{21}{13}=1.6153846$	$\frac{F_{13}}{F_{12}}$	$\frac{233}{144}=1.6180556$
$\frac{F_4}{F_3}$	$\frac{3}{2}=1.5$	$\frac{F_9}{F_8}$	$\frac{34}{21}=1.6190476$	$\frac{F_{14}}{F_{13}}$	$\frac{377}{233}=1.6180258$
$\frac{F_5}{F_4}$	$\frac{5}{3}=1.66\overline{6}$	$\frac{F_{10}}{F_9}$	$\frac{55}{34}=1.6176471$	$\frac{F_{15}}{F_{14}}$	$\frac{610}{377}=1.6180371$
$\frac{F_6}{F_5}$	$\frac{8}{5}=1.6$	$\frac{F_{11}}{F_{10}}$	$\frac{89}{55}=1.6181818$	$\frac{F_{16}}{F_{15}}$	$\frac{987}{610}=1.6180328$

Figure 2.18

Notice that as n gets larger, the approximation of each Fibonacci Ratio moves closer to the value of Φ. The ratios alternate, first smaller than Φ, then larger then Φ. This alternation seesaws back and forth, drawing closer and closer to the true value of Φ. Mathematically we say:

$$\lim_{n\to\infty}\frac{F_{n+1}}{F_n}=\Phi=\frac{1+\sqrt{5}}{2}\approx 1.61803398875...$$

This mathematical statement reads as follows: "As n approaches infinity, the limit of Fibonacci Ratios equals Phi, which is exactly $(1+\sqrt{5})/2$ or approximately 1.61803398875..."

Although we could write an entire book on the Golden Ratio (and many authors have done so), in order to make the important connection to poetry, we will confine our discussion to why it is thought of as the Divine Proportion. The key here is the word "proportion." When we say something is "proportional," we imply that it is pleasing to one of our senses. When we discuss construction involving the Golden Ratio, or Golden Proportion, whether we are discussing art, architecture, music, or poetry, we are using a proportion that will be aesthetically pleasing to our eye or ear.

Let's look first at the geometric origins of this aesthetic pattern. If we construct a rectangle whose length and width are respectively the long and short segments of Euclid's line (Figure 2.17), the result is known as the Golden Rectangle, shown in Figure 2.19.

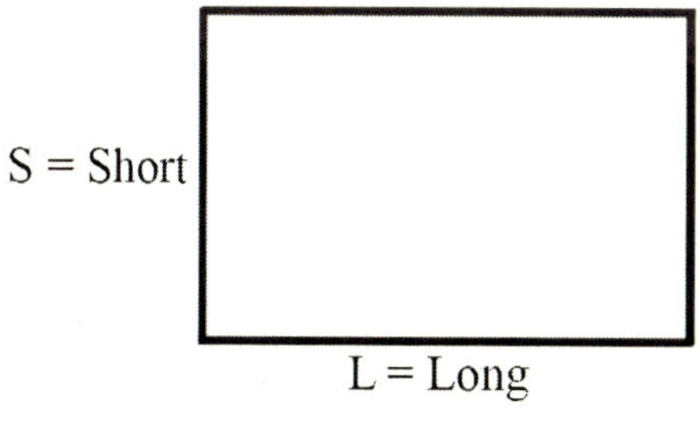

Figure 2.19

The rectangle whose sides fit the proportion

$$\frac{Long}{Short} = \frac{L}{S} = \frac{L+S}{L} = \Phi$$

is said to be the "most pleasing" in form of all rectangles. Claims have been made that it is found in many architectural examples, from the ancient Parthenon and Great Pyramid of Giza, to the more modern work of Le Corbusier. Close approximations of Golden Rectangles also may exist in painting and sculpture, including works by Leonardo da Vinci, Michelangelo, Georges Seurat and Piet Mondrian. Did the builders and artists make a conscious decision to include approximations of the Golden Rectangle in their work? In fact, are these rectangles close enough approximations to actually be considered "Golden?" As with the previous debate about Fibonacci Numbers, there are many questions and few definitive answers. However, some cubist and modern painters, sculptors, and architects have written about their appreciation for the Golden Mean and the application of Golden Ratios and Rectangles in their work.

If we manipulate a Golden Rectangle by removing square pieces, we discover another pattern. Let's begin with the Golden Rectangle drawn in Figure 2.20 and remove a square whose sides correspond to the short side of the rectangle. The rectangle that remains after the shaded square is removed is also a Golden Rectangle.

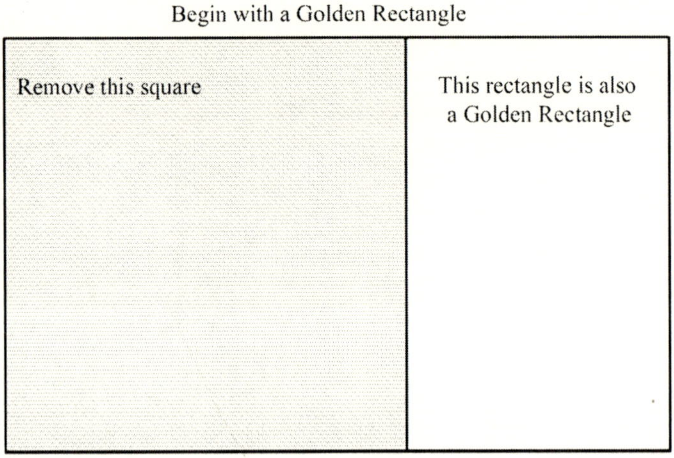

Figure 2.20

We can repeat this process through a number of steps, creating smaller and smaller Golden Rectangles. If we then connect the opposing corners of the squares, as shown in Figure 2.21, the resulting spiral is called the Logarithmic or Equiangular Spiral. This is the same spiral displayed in the nautilus and other shells, as well as in the seeds of sunflowers, bracts of pinecones (Figure 2.13), and the horns of many animals.

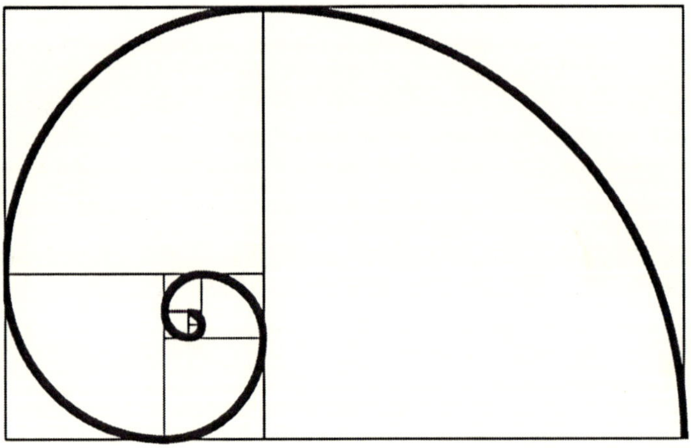

Figure 2.21

We can find many examples of poetry that refer to the beauty of the Golden Ratio, Golden Rectangle, or Logarithmic Spiral. For example, in the poem "Measures" Nadya Aisenberg tells us:

> [. . .] More than our days are numbered,
> children of Kronos. Speech, step, song.
> One rectangle so beautiful
> men call it golden,
> the Divine Proportion of the Parthenon [. . .]
>
> (Aisenberg 30)

And Oliver Wendell Holmes, in his poem "The Chambered Nautilus," uses the growth pattern of the nautilus shell as a metaphor for the experiences and spiritual growth in a human life. In this stanza, Holmes describes how the nautilus outgrows each chamber, building a new one in the pattern of the logarithmic spiral until it finally leaves the shell behind for a new home.

> Year after year beheld the silent toil
>   That spread his lustrous coil;
>   Still, as the spiral grew,
> He left the past year's dwelling for the new,
> Stole with soft step its shining archway through,
>   Built up its idle door,
> Stretched in his last-found home, and knew the old no more.
>
> (Holmes 149)

Pascal's Triangle, Fibonacci Numbers, and the Golden Ratio have come together in a variety of interesting intersections. These three ancient patterns remain a source of inspiration for both the serious and recreational mathematician, and the poet.

2.2 Traditional Poetic Forms: Patterns and Adaptations

As we've just seen, poets have been fascinated by the creative possibilities suggested by Fibonacci Numbers, the Golden Ratio, and the nautilus shell, but poetry also has a number of forms of its own. These forms, too, are recognized for their beauty, their structural properties, and their relative timelessness. When used in poetry, the word "form" refers to a set, replicable pattern where elements such as meter, rhyme, and number of lines come together to create a specific type of poem. When the adjectives "closed," "fixed," or "traditional" are used to describe these poetic forms, they are referring to important, defining qualities: the patterns of the forms were set long ago, and the forms themselves have been handed down and adapted over and over again by generations of writers. You probably won't be surprised to learn that some of these poetic forms have inherent mathematical properties. The patterns in fixed-form poetry are shaped by the repetition of words, sounds, even entire lines. Some patterns are readily apparent; other are less obvious. From among the many enduring traditional forms, we have chosen to focus on three: the sestina, sonnet, and villanelle. We have selected them for the variety of patterns they present and because of the hardiness with which they have survived. All of these forms are several hundred years old, and each originated in a language other than English. Through time, and in the hands of gifted poets, each form took on its own distinctive features. However, these forms are not static but have been continually subjected to adaptation and experimentation, as we'll see when we study each one in turn.

Sestina

Attributed to the twelfth-century Provençal poet Arnaut Daniel, the sestina is a good example of a poetic form constructed around a strong mathematical component. Rather than a repeating pattern of rhyme, the sestina relies on a pattern of six repeating *end words* to create its structural pattern. Let's see how the modern poet Elizabeth Bishop makes use of the sestina form in her poem "A Miracle for Breakfast."

Here's a suggestion: After you have read this poem through silently, read it again, aloud. Can you hear and follow the pattern of repeated words?

> At six o'clock we were waiting for coffee,
> waiting for coffee and the charitable crumb
> that was going to be served from a certain balcony,
> — like kings of old, or like a miracle.

It was still dark. One foot of the sun
steadied itself on a long ripple in the river.

The first ferry of the day had just crossed the river.
It was so cold we hoped that the coffee
would be very hot, seeing that the sun
was not going to warm us; and that the crumb
would be a loaf each, buttered, by a miracle.
At seven a man stepped out on the balcony.

He stood for a minute alone on the balcony
looking over our heads toward the river.
A servant handed him the makings of a miracle,
consisting of one lone cup of coffee
and one roll, which he proceeded to crumb,
his head, so to speak, in the clouds — along with the sun.

Was the man crazy? What under the sun
was he trying to do, up there on his balcony!
Each man received one rather hard crumb,
which some flicked scornfully into the river,
and, in a cup, one drop of coffee.
Some of us stood around, waiting for a miracle.

I can tell what I saw next; it was not a miracle.
A beautiful villa stood in the sun
and from its doors came the smell of hot coffee.
In front, a baroque white plaster balcony
added by the birds, who nest along the river,
— I saw it with one eye close to the crumb —

and galleries and marble chambers. My crumb
my mansion, made for me by a miracle,
through ages, by insects, birds, and the river
working the stone. Every day, in the sun,
at breakfast time I sit on my balcony
with my feet up, and drink gallons of coffee.

We licked up the crumb and swallowed the coffee.
A window across the river caught the sun
as if the miracle were working, on the wrong balcony.

(Bishop 18-19)

What are some of the qualities you observe as you read this poem? For one thing, you can probably see that it tells a story. When we look or listen closely, we can also observe that critical details of the story, "coffee," "crumb," "balcony," etc. are mentioned in each stanza. Yet with each repetition, there is a shift in focus, as the story takes another turn. Unlike in a ballad, where a story is often told from beginning to end, the narrative structure in a sestina is not always linear. The story may be presented in a way that seems illogical and dreamlike, as seen here, or it may be recursive, almost circular, a pattern seen in another widely-anthologized Elizabeth Bishop poem appropriately titled "Sestina."

The pattern that defines the sestina contributes to its characteristic narrative qualities. That pattern requires the six chosen end words to be repeated in a systematic way throughout the first six stanzas. In the three-line seventh stanza, called the *envoy*, all six words appear again, although the order may vary. Of course, the challenge for the poet in composing a sestina is to write a poem that uses the six repeating words in the correct order without becoming repetitious. Sometimes, the structural pattern of a sestina becomes virtually invisible as the reader enjoys the poem for its other aesthetic qualities; however, recognizing the sestina form and understanding the technical challenges it poses can certainly add to our appreciation of the poet's craft.

Mathematically, we would describe the pattern of the movement of the end words in the first six stanzas as a *permutation*. If we substitute symbols for the six repeating end words (coffee, crumb, balcony, etc.), we can observe the permutation, by following any given end word as it moves through the stanzas.

coffee	●●●●
crumb	❖❖❖
balcony	□□□
miracle	♏♏♏
sun	○○○
river	♒♒♒
river	♒♒♒
coffee	●●●●
sun	○○○
crumb	❖❖❖
miracle	♏♏♏
balcony	□□□
balcony	□□□
river	♒♒♒
miracle	♏♏♏
coffee	●●●●
crumb	❖❖❖
sun	○○○

sun	○○○
balcony	▢▢▢
crumb	❖❖❖
river	♒♒♒
coffee	●●●●
miracle	♏♏♏

miracle	♏♏♏
sun	○○○
coffee	●●●●
balcony	▢▢▢
river	♒♒♒
crumb	❖❖❖

crumb	❖❖❖
miracle	♏♏♏
river	♒♒♒
sun	○○○
balcony	▢▢▢
coffee	●●●●

crumb coffee	❖❖❖	●●●●
river sun	♒♒♒	○○○
miracle balcony	♏♏♏	▢▢▢

Figure 2.22

When we arrange the stanzas in columns, as shown below, the pattern becomes even more apparent.

Stanza 1	Stanza 2	Stanza 3	Stanza 4	Stanza 5	Stanza 6
●●●●	♒♒♒	▢▢▢	○○○	♏♏♏	❖❖❖
❖❖❖	●●●●	♒♒♒	▢▢▢	○○○	♏♏♏
▢▢▢	○○○	♏♏♏	❖❖❖	●●●●	♒♒♒
♏♏♏	❖❖❖	●●●●	♒♒♒	▢▢▢	○○○
○○○	♏♏♏	❖❖❖	●●●●	♒♒♒	▢▢▢
♒♒♒	▢▢▢	○○○	♏♏♏	❖❖❖	●●●●

Figure 2.23

Numbering the words in each column as 1, 2, . . . 6, the mathematician would describe the permutation as: (1, 2, 4, 5, 3, 6), that is, the word at the end of line 1 moves first to the end of line 2, then to line 4, next to line 5, next to line 3, and finally to line 6. If you begin with any column and follow a symbol or a single word, coffee (●●●●) for example, you will see how the words move through the permutation.

Although using a permutation is critical when writing a sestina, permutations themselves are firmly grounded in the domain of the mathematician. This link between poetry and mathematics may be extended even further by comparing the shifting of words in a sestina to the shift of digits in a *cyclic number*[5].

Mathematically, a cyclic number is found by first finding a prime number n such that its reciprocal has n-1 digits in its repeating portion. The smallest cyclic number is the 6-digit number 142857 and it is related to the prime number 7. The decimal representation for 1/7 is 0.142857142857... The repeating portion is made up of 6 digits. If we multiply the number 142857 by the Natural Numbers 1 through 6, the answers are each a permutation of the same 6 digits in the same cycle order. No matter where the number "1" occurs in any of the 6 answers, if we read to the right edge of the number and then start back again at the left edge, the digits read "1 4 2 8 5 7."

1 x 142847 = 142857
2 x 142857 = 285714
3 x 142857 = 428571
4 x 142857 = 571428
5 x 142857 = 714285
6 x 142857 = 857142

For this cyclic number the permutation of "1" is (1, 5, 6, 3, 2, 4), that is "1" goes from leftmost position (call it 1st position) to the 5th position, then moves to the 6th position, and so on. We have made the numeral "1" larger in each line to make it easier to follow the shift.

The sestina is not as well known as the limerick or the sonnet, yet it has intrigued poets for centuries, posing as it does a technical and creative challenge. Over the years, poets have experimented with the use of rhyme in the poem and with double sestinas (using twelve repeating end words in twelve stanzas and a final envoy). Contemporary poets have used the sestina to write about subjects ranging from the bleakness of winter in Rochester, New York ("Sestina d'Inverno" by Anthony Hecht) to the cartoon character of Popeye,

[5] Further explanation of cyclic numbers is found in Ch. 10 of Martin Gardner's *Mathematical Circus*. Washington DC: Mathematical Assoc. of America, 1992, 111-122.

featured in the rhymed sestina "Farm Implements and Rutabagas in a Landscape" by John Ashbery.

If writing with a six-part permutation seems tough, and if rhyming adds another element to the challenge, imagine taking it a step further and writing a rhymed *double sestina*, as done by Algernon Charles Swinburne in "The Complaint of Lisa." We've put the first pair of rhyming end-words, *breath* and *death*, in capital letters so that you may follow their movement through the poem. (The other rhyming, repeating words you will find here are "her" and "sunflower," "way" and "day," "sun" and "done," "bed" and "dead," and "thee" and "me.")

There is no woman living that draws BREATH
So sad as I, though all things sadden her.
There is not one upon life's weariest way
Who is weary as I am weary of all but DEATH.
Toward whom I look as looks the sunflower
All day with all his whole soul toward the sun;
While in the sun's sight I make moan all day,
And all night on my sleepless maiden bed.
Weep and call out on death, O Love, and thee,
That thou or he would take me to the dead.
And know not what thing evil I have done
That life should lay such heavy hand on me.

Alas! Love, what is this thou wouldst with me?
What honor shalt thou have to quench my BREATH,
Or what shall my heart broken profit thee?
O Love, O great god Love, what have I done,
That thou shouldst hunger so after my DEATH?
My heart is harmless as my life's first day:
Seek out some false fair woman, and plague her
Till her tears even as my tears fill her bed:
I am the least flower in thy flowery way,
But till my time be come that I be dead,
Let me live out my flower-time in the sun,
Though my leaves shut before the sunflower.

O Love, Love, Love, the kingly sunflower!
Shall he the sun hath looked on look on me,
That live down here in shade, out of the sun,
Here living in the sorrow and shadow of DEATH?
Shall he that feeds his heart full of the day

Care to give mine eyes light, or my lips BREATH?
Because she loves him shall my lord love her
Who is as a worm in my lord's kingly way?
I shall not see him or know him alive or dead;
But thou, I know thee, O Love, and pray to thee
That in brief while my brief life-days be done,
And the worm quickly make my marriage-bed.

For underground there is no sleepless bed:
But here since I beheld my sunflower
These eyes have slept not, seeing all night and day
His sunlike eyes, and face fronting the sun.
Wherefore if anywhere be any DEATH,
I fain would find and fold him fast to me,
That I may sleep with the world's eldest dead,
With her that died seven centuries since, and her
That went last night down the night-wandering way.
For this is sleep indeed, when labor is done,
Without love, without dreams, and without BREATH,
And without thought, O name unnamed! of thee.

Ah! but, forgetting all things, shall I thee?
Wilt thou not be as now about my bed
There underground as here before the sun?
Shall not thy vision vex me alive and dead,
Thy moving vision without form or BREATH?
I read long since the bitter tale of her
Who read the tale of Launcelot on a day,
And died, and had no quiet after DEATH,
But was moved ever along a weary way,
Lost with her love in the underworld; ah me,
O my king, O my lordly sunflower,
Would God to me, too, such a thing were done!

But if such sweet and bitter things be done,
Then, flying from life, I shall not fly from thee.
For in that living world without a sun
Thy vision will lay hold upon me dead,
And meet and mock me, and mar my peace in DEATH.
Yet if being wroth God had such pity on her,
Who was a sinner and foolish in her day,
That even in hell they twain should breathe one BREATH,

Why should he not in some wise pity me?
So if I sleep not in my soft strait bed,
I may look up and see my sunflower
As he the sun, in some divine strange way.

O poor my heart, well knowest thou in what way
This sore sweet evil unto us was done.
For on a holy and a heavy day
I was arisen out of my still small bed
To see the knights tilt, and one said to me
"The king," and seeing him, somewhat stopped my BREATH,
And if the girl spake more, I heard not her,
For only I saw what I shall see when dead.
A kingly flower of knights, a sunflower,
That shone against the sunlight like the sun,
And like a fire, O heart, consuming thee,
The fire of love that lights the pyre of DEATH.

Howbeit I shall not die an evil DEATH
Who have loved in such a sad and sinless way,
That this my love, lord, was no shame to thee.
So when mine eyes are shut against the sun,
O my soul's sun, O the world's sunflower,
Thou nor no man will quite despise me dead.
And dying I pray with all my low last BREATH
That thy whole life may be as was that day,
That feast-day that made trothplight death and me,
Giving the world light of thy great deeds done;
And that fair face brightening thy bridal bed,
That God be good as God hath been to her.

That all things goodly and glad remain with her,
All things that make glad life and goodly DEATH;
That as a bee sucks from a sunflower
Honey, when summer draws delighted BREATH,
Her soul may drink of thy soul in like way,
And love make life a fruitful marriage-bed
Where day may bring forth fruits of joy to day
And night to night till days and nights be dead.
And as she gives light of her love to thee,
Give thou to her the old glory of days long done;

And either give some heat of light to me,
To warm me where I sleep without the sun.

O sunflower made drunken with the sun,
O knight whose lady's heart draws thine to her,
Great king, glad lover, I have a word to thee.
There is a weed lives out of the sun's way,
Hid from the heat deep in the meadow's bed,
That swoons and whitens at the wind's least BREATH,
A flower star-shaped, that all a summer day
Will gaze her soul out on the sunflower
For very love till twilight finds her dead.
But the great sunflower heeds not her poor DEATH,
Knows not when all her loving life is done;
And so much knows my lord the king of me.

Ay, all day long he has no eye for me;
With golden eye following the golden sun
From rose-colored to purple-pillowed bed,
From birthplace to the flame-lit place of DEATH,
From eastern end to western of his way.
So mine eye follows thee, my sunflower,
So the white star-flower turns and yearns to thee,
The sick weak weed, not well alive or dead,
Trod under foot if any pass by her,
Pale, without color of summer or summer BREATH
In the shrunk shuddering petals, that have done
No work but love, and die before the day.

But thou, to-day, to-morrow, and every day,
Be glad and great, O love whose love slays me.
Thy fervent flower made fruitful from the sun
Shall drop its golden seed in the world's way,
That all men thereof nourished shall praise thee
For grain and flower and fruit of works well done;
Till thy shed seed, O shining sunflower,
Bring forth such growth of the world's garden-bed
As like the sun shall outlive age and DEATH.
And yet I would thine heart had heed of her
Who loves thee alive; but not till she be dead.
Come, Love, then, quickly, and take her utmost BREATH.

Song, speak for me who am dumb as are the dead;
From my sad bed of tears I send forth thee,
To fly all day from sun's birth to sun's DEATH
Down the sun's way after the flying sun,
For love of her that gave thee wings and BREATH
Ere day be done, to seek the sunflower.

(Swinburne 44-49)

If we consider only the first twelve stanzas, without the envoy, and follow the same two words, *breath* and *death*, with the symbols & and ⌘, respectively, their movement throughout the poem's twelve stanzas looks like this:

◄——— Stanzas ———►

Line	1	2	3	4	5	6	7	8	9	10	11	12
1	&							⌘				
2		&							⌘			
3												
4	⌘		⌘						&		⌘	
5		⌘		⌘	&	⌘						
6			&				&			&		
7								&				
8					⌘	&						
9												⌘
10										⌘	&	
11				&								
12							⌘					&

Figure 2.24

To maintain a system of repeated end words, while limiting himself to using six rhyming pairs for those end words, Swinburne needed to create his own pattern for shifting the twelve key words through the stanzas. Although this poem does not have a repeating pattern as clear as the permutation in a single sestina, we can observe some similarities to the 6-stanza sestina. For example, the last end word in each stanza of the double sestina becomes the first end word in the new stanza, just as it does in the single sestina; similarly, the first end word of any given stanza becomes the second end word of the following stanza in both the single and double sestinas. Beyond that, however, there is a great deal of variation in the pattern each word follows in Swinburne's poem. Some words appear in the same place in several stanzas ("sun" appears four times as the end word of the third line) and other words never appear in a particular line (for instance, "breath" never appears as the end word in the third or ninth line).

If you enjoy the challenge of thinking creatively within a complicated framework, you might like to try your hand at writing a sestina, whether rhymed or unrhymed, single or double. Like the challenge of finding creative ways to checkmate your opponent in a game of chess, the challenge of telling a story within the constraints of this tricky fixed form can be both absorbing and exhilarating. Perhaps, a bit frustrating, too!

Sonnet

In comparison to the rhyming double sestina, the traditional sonnet may look deceptively simple. The sonnet is certainly one of the best known forms in English, although its origins are Italian. The word comes from *sonetto*, meaning "little song," and refers to a relatively short poetic form made popular by the fourteenth-century Italian poet Francesco Petrarch. The patterns at work in a sonnet are created by a combination of rhyme, meter, and the grouping of lines. The Italian sonnet usually consists of 14 lines and is divided into two sections: the *octave* (the first 8 lines) and the *sestet* (the concluding 6 lines). Often the octave is used to establish the setting or central idea of the poem, and the sestet then extends, complicates, or resolves that idea. Here is an example of the Italian form, "Zephyr returns, and scatters everywhere," a sonnet written by Petrarch and translated into English by Morris Bishop, who maintained the rhyme scheme of the original (*abababab cdcdcd*). There are a number of variations to the rhyme scheme of the Italian sonnet, but what distinguishes it unmistakably is the division into two integral units of 8 and 6 lines.

The octave section of this poem includes many allusions or references to ancient mythology, as Petrarch describes the return of spring, beginning with the appearance of Zephyr, the western wind. The language describes a bountiful, joyous season; however, with the first word of the sestet, line 9 ("But"), the poet signals that, for the speaker, the spring brings not joy but "heavy sighs." This shift in focus is characteristic of the sestet. It is referred to as the *volta*, or "turn."

Zephyr returns, and scatters everywhere — a
New flowers and grass, and company does bring, — b
Procne and Philomel, in sweet despair, — a
And all the tender colors of the Spring. — b
Never were fields so glad, nor skies so fair; — a
And Jove exults in Venus' prospering. — b
Love is in all the water, earth, and air, — a
And love possesses every living thing. — b
But to me only heavy sighs return — c
For her who carried in her little hand — d
My heart's key to her heavenly sojourn. — c
The birds sing loud above the flowering land; — d
Ladies are gracious now. -- Where deserts burn — c
The beasts still prowl on the ungreening sand. — d

(Petrarch 205)

Now let's look at an example of a sonnet written in English, but using the Italian form. The poem, "Euclid alone has looked on Beauty bare," pays tribute to the ancient Greek mathematician. Readers who are not mathematicians may not be familiar with the long tradition of using poetry to celebrate advances in mathematics. Many such poems have been written by mathematicians themselves. Here, though, it's Edna St. Vincent Millay, an American woman poet from the first half of the twentieth century who is writing a sonnet to celebrate a landmark in mathematical thought.

Euclid alone has looked on Beauty bare. — a
Let all who prate of Beauty hold their peace, — b
And lay them prone upon the earth and cease — b
To ponder on themselves, the while they stare — a
At nothing, intricately drawn nowhere — a
In shapes of shifting lineage; let geese — b
Gabble and hiss, but heroes seek release — b
From dusty bondage into luminous air. — a
O blinding hour, O holy, terrible day, — c
When first the shaft into his vision shone — d
Of light anatomized! Euclid alone — d
Has looked on Beauty bare. Fortunate they — c
Who, though once only and then but far away, — c
Have heard her massive sandal set on stone. — d

(Millay *Harp-Weaver* 70)

Although Millay chose to write in the Italian form, there are two alternative forms of the sonnet that have developed in English. Both are named for the writers who popularized them: William Shakespeare and Edmund Spenser. Of the two, the Shakespearean Sonnet, also known as the English Sonnet, is more familiar. Like its Italian forebear, the English Sonnet has 14 lines. It differs, however, in the way the rhyme scheme and the poem's ideas are arranged within those 14 lines. Rather than being divided into two sections of 8 and 6 lines each, the lines of the English Sonnet are arranged in three sets of 4 lines, known as *quatrains*, and a final *couplet*, or pair of rhymed lines. Each quatrain has a rhyme scheme of alternating rhymed lines. The diagram in Figure 2.25 summarizes the differences in structure between the Petrarchan or Italian Sonnet and the Shakespearean or English Sonnet. It also shows two variations of the Italian Sonnet rhyme scheme. It's important to remember that these simple models of structure and rhyming patterns cannot begin to reveal the subtlety or complexity of ideas that may be developed in 14 lines.

Petrarchan Sonnet

	Variation 1	Variation 2
Octave	a	a
	b	b
	a	b
	b	a
	a	a
	b	b
	a	b
	b	a
Sestet	c	c
	d	d
	c	e
	d	c
	c	d
	d	e

Shakespearean Sonnet

Quatrain	a
	b
	a
	b
Quatrain	c
	d
	c
	d
Quatrain	e
	f
	e
	f
Couplet	g
	g

Figure 2.25

The following sonnet by William Shakespeare shows beautifully how the three quatrains work together. The rhymed couplet, in turn, allows the poet to step back from the poem, sometimes to be ironic or disclosive, or to put the theme of the poem in a larger context. We will return to Sonnet 73 in Chapter 4 when we consider how a familiarity with fractals might influence our reading of this poem.

That time of year thou mayst in me behold a
When yellow leaves, or none, or few, do hang b
Upon those boughs which shake against the cold, a
Bare ruined choirs where late the sweet birds sang. b
In me thou seest the twilight of such day c
As after sunset fadeth in the west, d
Which by and by black night doth take away, c
Death's second self that seals up all in rest. d
In me thou seest the glowing of such fire e
That on the ashes of his youth doth lie, f
As the deathbed whereon it must expire, e
Consumed with that which it was nourished by. f
This thou perceiv'st, which makes thy love more strong g
To love that well which thou must leave ere long. g

(Shakespeare 1465)

Poets have demonstrated repeatedly that despite its longevity, the sonnet can still be a powerfully expressive form. "If We Must Die,"[6] written by the Jamaican poet Claude McKay during the 1919 U.S. race riots, infuses the traditional sonnet form with all of the courage and force needed to confront an enemy. Years later, the poem was read publicly by Winston Churchill as a rallying cry in World War II.

If we must die, let it not be like hogs
Hunted and penned in an inglorious spot,
While round us bark the mad and hungry dogs,
Making their mock at our accursed lot.
If we must die, O let us nobly die,
So that our precious blood may not be shed
In vain; then even the monsters we defy
Shall be constrained to honor us though dead!
O kinsmen! we must meet the common foe!
Though far outnumbered let us show us brave,
And for their thousand blows deal one deathblow!
What though before us lies the open grave?

---

[6] In discussing his use of lyric poetry to express the impassioned ideas found in "If We Must Die," McKay asserts that, "If we [he and his colleagues on the *Liberator* magazine, where the poem was first published] were a rebel group because we had faith that human life might be richer, by the same token we believed in the highest standards of creative work." McKay, Claude. *The Passion of Claude McKay: Selected Prose and Poetry 1912 - 1948*, Ed. Wayne Cooper. New York: Schocken Books, 1973, 134.

Like men we'll face the murderous, cowardly pack,
Pressed to the wall, dying, but fighting back!

(McKay 124)

This moving *cri de coeur* is written in a traditional "closed" form, yet the form becomes a virtually transparent bulwark, supporting the poem's challenge and its loud, confident claims.

The "rules" or conventions of the sonnet form appear fairly straightforward when set down in a chart such as the one above, but poets have found countless ways to work within and challenge these constraints. The contemporary poet Gerald Stern has a written collection of poems titled *American Sonnets* comprising fifty-nine sonnets of varying lengths: the book includes sonnets of 18, 20, even 22 lines, but none is 14 lines long.

Here are a few other examples of experiments with the form: a sonnet whose language and use of punctuation is uniquely challenging; one composed of symbols rather than words; and another that almost entirely disregards the conventions of the sonnet.

It is fourteen lines in length and has a recognizable rhyme scheme, but E.E. Cummings' sonnet titled ")when what hugs stopping earth than silent is" challenges the reader on several levels, beginning with an idiosyncratic use of punctuation, namely, the use of the *closing* parenthesis at the *beginning* of the first line.

)when what hugs stopping earth than silent is
more silent than more than much more is or
total sun oceaning than any this
tear jumping from each most least eye of star

and without was if minus and shall be
immeasurable happenless unnow
shuts more than open could that every tree
or than all life more death begins to grow

end's ending then these dolls of joy and grief
these recent memories of future dream
these perhaps who have lost their shadows if
which did not do the losing spectres mime

until out of merely not nothing comes
only one snowflake(and we speak our names

(*50 Poems, in Poems 1923-1954,* 361)

When we get to the poem's last line, we find the opening half of the parenthesis, and are thus drawn into reading the poem as a circling of ideas that actually begins and ends in the middle of the *last* line.

In the following unusual looking poem, "Moon shot," Mary Ellen Solt, one of the "concrete poets," whose work we will explore in more detail in the Chapter 3, displays an awareness of her role as an innovator within a long-standing poetic tradition when she uses symbols borrowed from science to create the sonnet's octave and sestet. Solt herself has said, "[the poem was] made by copying the scientists' symbols on the first photos of the moon in *The New York Times*: there were exactly fourteen 'lines' with five 'accents.' We have not been able to address the moon in a sonnet successfully since the Renaissance. Admitting its new scientific content made it possible to do so again. The moon is a different object today [...] the poem is both a spoof of old forms and a statement about the necessity for new" (Solt 307).

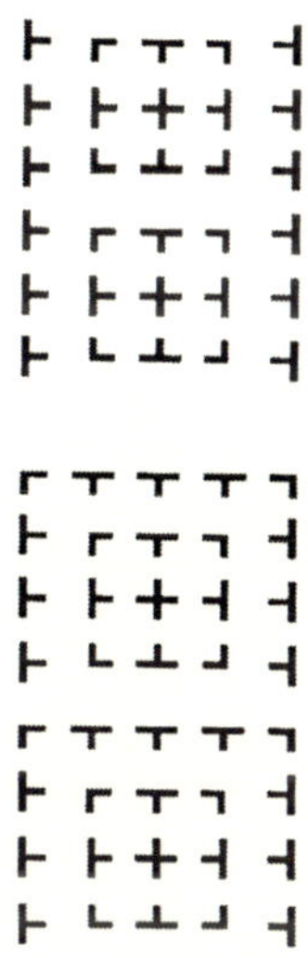

(Solt 242)

Sometimes, experiments to the traditional sonnet form virtually overthrow all of our expectations. Charles North's "Sonnet" brings into question every assumption we might have about the sonnet's meter, rhyme, and line length.

The tone poem left the door open.
Well, *close* it.
It doesn't stay. It reminds me of

Elizabethan plays where eyes,
especially the tragically blinded ones, are "jelly."
It has a center with a circumference loosely attached.
The ideas about social wastefulness
smeared over individual needs.

Since the ideas about wastefulness
are smeared over their objects,
the tone is everywhere.
It expresses its reluctance as virtue.
It is reluctant to intrude, like minds into
the fleetingness they concede.

(*Best American Poetry 2002*, 124)

We'll bring this examination of the sonnet to a close with a sophisticated contemporary poem that re-makes the traditional Shakespearean sonnet with freshness and irony, while retaining many of the sonnet's formal properties.

Beginning with the title, "Sequence Sonnet," Kelly Cherry creates a play-on-words by borrowing a term that usually refers to a series of thematically-linked sonnets. Because Cherry's poem follows the "sequence" of love all the way from courtship to loss, it serves as an ironic reference to the idea of sequence, here truncating the course of an entire relationship into 14 lines; it also challenges the idea of the sonnet as a "love poem," a role it often serves.

This poem's re-making of the sonnet extends to an innovative use of sound, as Cherry broadens the Shakespearean sonnet's rhyme scheme to allow less than perfect end-rhymes. The rhyming couplet at the end is easily recognizable, but other subtle uses of rhyme such as the pairings of "house" and "compose," or "time" and "tame," give the impression of layered, echoing sounds, rather than the precise repetitions of exact rhyme, such as we find in Shakespeare's "Sonnet 73" or Claude McKay's "If We Must Die."

I will bring bouquets to his house,
Volumes of verse and Beethoven's Ninth,
Seventh, Fifth, Third and Eighth.
Gaining confidence, I will compose
Sonnets on the subject of masculine beauty
As a direct function of time.
When his face has become a mockery
Of the one I fell for, I will tame
My laughter like a savage cat
Washing itself on the high cliff.
In our passage from love to the last, wild, selfish grief

Which knows nothing, I will conduct
This gentle man through the garden of courtship.

(Cherry 20)

Villanelle

The villanelle is another traditional poetic form that originated in Italy. Rooted in folk song, the villanelle form has remained virtually the same since being set down by the French poet Jean Passerat in the 1500s. In the villanelle, the dominant pattern of the poem is not created by repeating individual words, nor with rhymed blocks or units of thought, but by the incremental development of ideas that results when *entire lines are repeated.* A consistent *aba* rhyme scheme throughout the poem adds to its sense of unity. The pattern within the form is quite simple: The first five stanzas consist of three lines, and the sixth stanza is a quatrain. The first and third lines of the first stanza are repeated throughout the poem, taking turns as the final line in stanzas two, three, four, and five, and then coming together again as the last two lines of the final stanza. The two repeating lines rhyme with one another, and rhyme with the first line of each stanza as well (these are marked below as A1, A2, and a). The second lines of each stanza also rhyme with one another; thus, throughout the poem, there are only two different sounds to the end rhymes. This modern villanelle, "The House on the Hill" by Edwin Arlington Robinson, uses the repeating lines very effectively to create a growing sense of emotional intensity and loss.

They are all gone away,	A
The House is shut and still,	b
There is nothing more to say.	A2
Through broken walls and gray	a
The winds blow bleak and shrill:	b
They are all gone away.	A1
Nor is there one to-day	a
To speak them good or ill:	b
There is nothing more to say.	A2
Why is it then we stray	a
Around the sunken sill?	b
They are all gone away,	A1

And our poor fancy-play a
For them is wasted skill: b
There is nothing more to say. A2

There is ruin and decay a
In the House on the Hill: b
They are all gone away, A1
There is nothing more to say. A2
(Robinson 81-82)

With its two repeating lines and overall *aba* rhyme scheme, the patterns in the villanelle are easy to recognize, but more difficult to create. As when the poet is writing a sestina, one major challenge in writing the villanelle comes in selecting elements that will contribute to the poem and — through repetition — add depth and interest, rather than redundancy. The distribution of repeated elements becomes apparent when those lines are capitalized:

THEY ARE ALL GONE AWAY,
The House is shut and still,
THERE IS NOTHING MORE TO SAY.

Through broken walls and gray
The winds blow bleak and shrill:
THEY ARE ALL GONE AWAY,

Nor is there one to-day
To speak them good or ill:
THERE IS NOTHING MORE TO SAY.

Why is it then we stray
Around the sunken sill?
THEY ARE ALL GONE AWAY,

And our poor fancy-play
For them is wasted skill:
THERE IS NOTHING MORE TO SAY.

There is ruin and decay
In the House on the Hill:
THEY ARE ALL GONE AWAY,
THERE IS NOTHING MORE TO SAY.

As a final example of experiments in closed form, "Icebreaker" by Chris Wiltz uses the repetitions of the villanelle in an entirely different way, creating the sense of possibility, rather than closure. Reading the poem aloud, you can hear how its repeating elements convey the speaker's hesitation and eagerness, as we eavesdrop on one side of a conversation:

Hey, I really like your ___ because –
Wait, I just wanna say:
No, I mean ... what I meant to say was ...

Your eyes ... I mean your hair was –
__ No wait, lemme start over okay?
Hey, I really like your ___ because –

I-was-standing-over-there ... and I swear I don't have a buzz
But I found-you-intriguing-in-some-way
No, I mean ... what I meant to say was ...

Do-you ... work out during the day?
(I'd say something about you running but that's kinda cliché.)
Hey, I really like your ___ because –

Would you like my drink / a drink? ... I mean, can I buy you a drink because ...
No I'm not trying to take advantage I just think___ I saw you looking this way.
No, I mean ... what I meant to say was ...

I don't suppose you read Maxim because___
There was this really good article (I don't remember) from the other day ... anyway
Hey, I really like your ___ because –
No, I mean ... what I meant to say was ...

(Wiltz)

The poems presented here demonstrate how patterns of repetition, often influenced by music or mathematics, have created poetic forms that are pleasing and enduring. Rather than hindering the poet's expressiveness or creativity, the conventions that define these forms provide the opportunity for experimentation. Attracted by the interplay between fixed form and creativity, many contemporary writers have made dynamic contributions to this poetic tradition.

Chapter 3 – Patterns of Shape

The topic of this chapter is patterns of shape. We find shapes throughout the natural world and in the tools, structures, and art made by human hands. In mathematics, shapes are usually thought of as figures. They may be regular or irregular; they may exist in two-dimensions or three; they may be repeated or placed in patterns that are described as symmetrical or asymmetrical. Mathematicians identify, classify, and manipulate shapes. Artists, including M. C. Escher, have shown us that mathematical shapes may be the basis for designs that challenge the mind as well as the eye. For poets, the shape of a poem — what it looks like on the page — is sometimes just as important as or inextricably linked to the poem's ideas. In fact, there is a specific type of poetry known as "shaped" or "pattern" poetry. We can find both ancient and contemporary examples of these poems, cut into stone, composed letter-by-letter on manual typewriters, or made by computers. Whether conscious of it or not, poets are also influenced by mathematical ideas about shapes when they apply principles of symmetry to their work by creating or breaking symmetry in the arrangement of words or ideas. Of course, concepts drawn from mathematics may be explored for their suggested or implied meanings too, as we'll see in the final poem in this chapter, where a scientific theory based on symmetry and asymmetry becomes a metaphor for the creative work of the poet.

Let's begin with mathematical shapes.

3.1 Mathematical Shapes

Our ancestors observed repetition and regularity in shapes, as well as in numbers. Certain rocks made better weapons than others; mountains had pointed tops; leaves of plants often shared a basic form; some shelters built from stones or sticks were more stable than others. The earliest concepts of geometric shape were formed long before the words *circle, square*, and *triangle* had been invented. When ancient peoples made weapons, created vessels to hold food and water, or decorated their surroundings, the shapes they chose were often not unique to their culture. Triangle-shaped arrow heads left by early Native American tribes resemble those left by Roman conquerors in the British Isles; circular woven baskets from ancient Greece are akin to those made by primitive tribes in Africa; and circular suns and triangular mountains appear in cave art of Spain and France, as well as rock-wall petroglyphs in New Mexico and Utah.

Plane Figures

From ancient times people have looked at the two-dimensional and three-dimensional shapes that surround us, searching for similarities and differences. Mathematicians have always delighted in finding the patterns, both obvious and hidden, that allowed shapes to be named, categorized, and replicated. It is not sufficient to know that triangles can be classified either by the length of their sides or by the size of their angles. The ability to reproduce different triangles, through a clear set of instructions, is equally important. The visual concept of recognizing shape has to be supplemented with the tactile act of drawing the figure. Some of the earliest graphic artists were mathematicians who set out the rules of geometric construction for the shapes known as "plane figures."

What are plane figures? Plane figures, including squares, rectangles, and triangles, are two-dimensions shapes that may be drawn on a flat piece of paper or plane surface. There are many ways to classify plane figures, such as by the number of sides, by the types of angles, and by their construction.

Before starting school, many young children play with shape-sorting toys, such as a cube with cut-outs for inserting stars, squares, and circles; ordering toys, such as a spindle with different sized rings; and books that link shapes and size with simple rhymes. In grade school, students learn the names of circles, triangles, squares, and rectangles and how to classify them. Later they learn about quadrilaterals, the general term for four-sided figures, and start to recognize differences between parallelograms, which have two pairs of parallel sides, and trapezoids, which have only one pair of parallel sides.

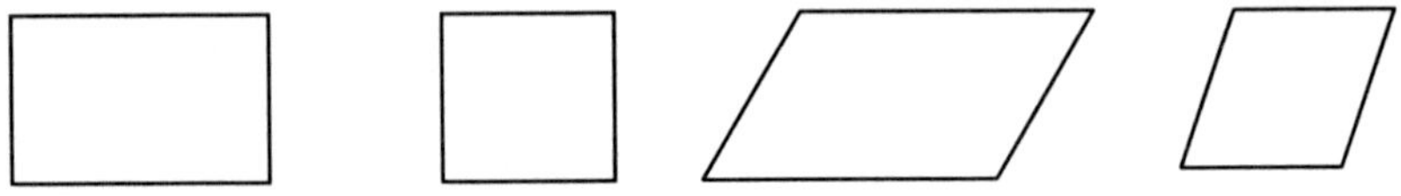

Parallelograms

Figure 3.1

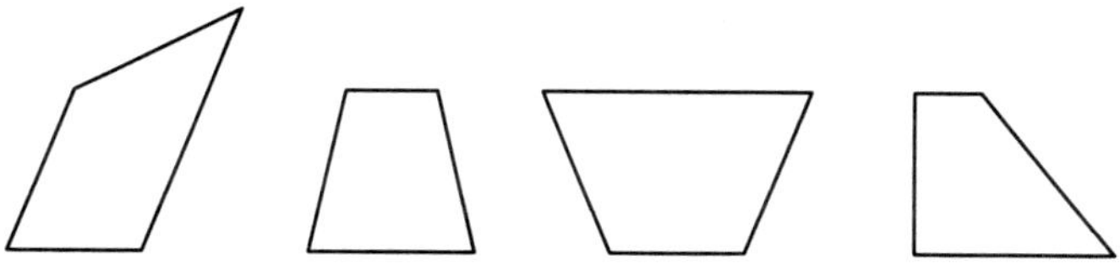

Trapezoids

Figure 3.2

In middle school or high school, students study plane geometry, learning more sophisticated concepts of point, line, and angle. The shapes they study are referred to as two-dimensional, plane, or closed figures. Often groups of plane figures having the same number of sides are studied in isolation to find the patterns within each group. For instance, triangles can be classified either by sides or by interior angles, in either case resulting in three categories:

By side: Equilateral — all sides of the same length
Isosceles — at least two sides of the same length
Scalene — all three sides of different lengths

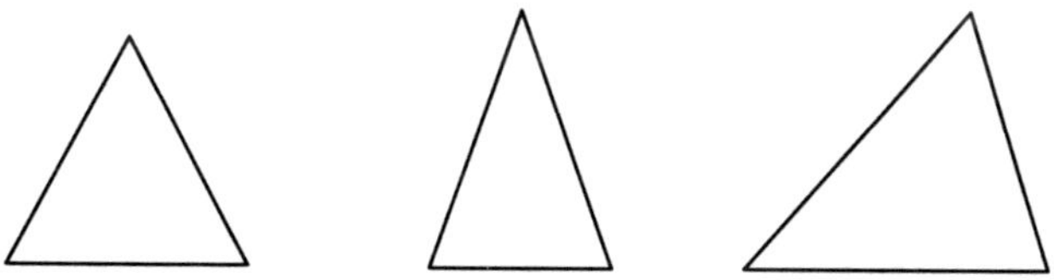

Equilateral Isosceles Scalene

Figure 3.3

By angle: Acute — all angles less than 90°
Right — one angle of exactly 90° and two angles less than 90°
Obtuse — one angle greater than 90° and two less than 90°

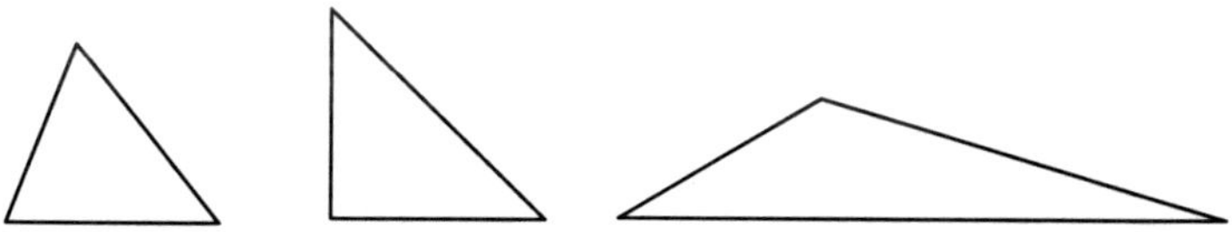

Acute Right Obtuse

Figure 3.4

Why do mathematicians examine the patterns within geometric shape groups? One answer is that this type of pattern analysis may lead new insights about the subject. Each of the triangles in Figure 3.3 exhibits other properties specific to that type of triangle. For instance, in an equilateral triangle all three angles and all three sides are equal; in an isosceles triangle the two angles opposite the equal sides are the same, but the third angle is different in size; finally, in the scalene triangle, not only are the three angles of different size, but the lengths of the sides are all different, too, with the largest angle opposite the longest side and the smallest angle opposite the shortest side. To understand geometry, a mathematician often looks for a pattern, classifies the pattern into subcategories, and then finds new patterns within these subcategories. It is a familiar problem-solving technique that mathematicians use in analyzing new mathematical situations. In Chapter 5, we will return to this type of geometric pattern-analysis when we discuss the concept of proof.

Gerald Lynton Kaufman in his book *Geo-metric Verse* playfully exploits the important distinctions between acute and obtuse triangles in forming his shaped poem "Sigh-Angles." Not only is the short verse written about triangles, but the words themselves form two triangles, one acute and the other obtuse, whose sides will never coincide. Later in this chapter, we'll have more to say about shaped poetry.

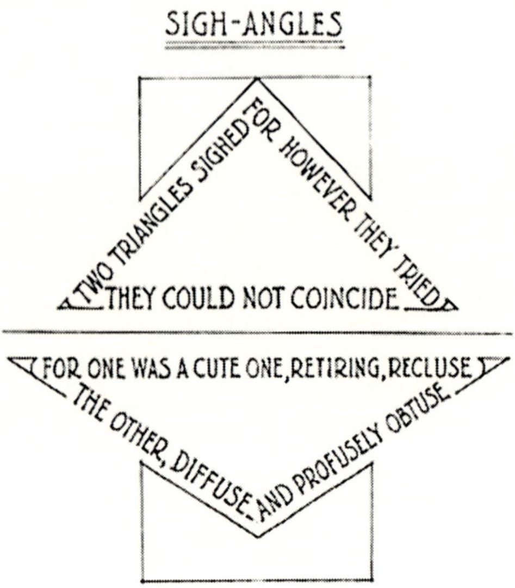

(Kaufman 7)

As we continue classifying shape patterns, it's important to know what is meant by a *regular* plane figure. These are closed figures whose sides have equal length and whose interior angles are all of the same size. All three characteristics must be true for a figure to be classified as regular. Some examples of regular and irregular plane figures illustrate this concept in Figure 3.5.

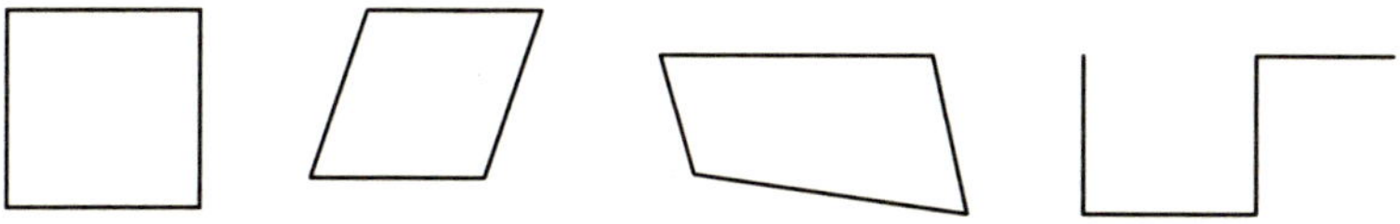

Regular	Irregular	Irregular	Irregular
Closed	Closed	Closed	Open
4 Sides Equal	4 Sides Equal	Sides Not Equal	4 Sides Equal
4 Angles Equal	Angles Not Equal	Angles Not Equal	Angles Equal

Figure 3.5

The table in Figure 3.6 shows the names and pictures of some of the *regular* closed plane figures and corresponding *irregular* closed plane figures having the same number of sides. All the shapes in the table are called polygons, from the Greek words "polys" meaning "many" and "gonia" meaning "angle."

Sides	Name of Shape	Regular Plane Figure	Irregular Plane Figure
3	Triangle	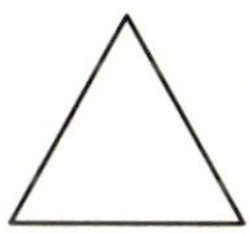	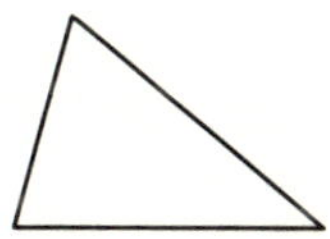
4	Quadrilateral		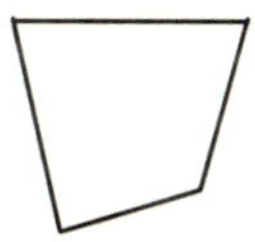
5	Pentagon	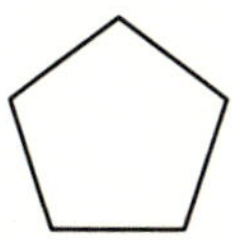	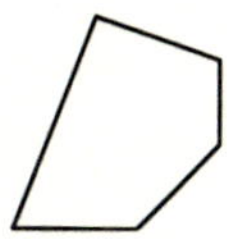

6	Hexagon	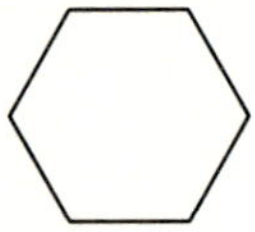	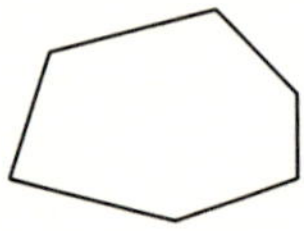
7	Heptagon	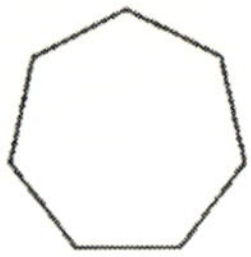	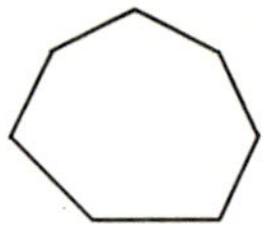
8	Octagon	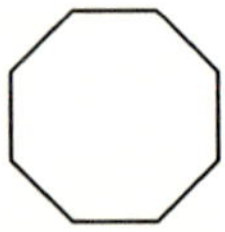	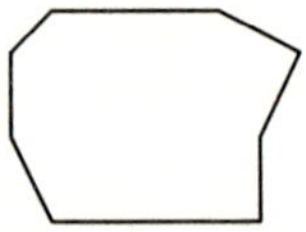

Figure 3.6

We will end the section on plane figures with an homage to two shapes considered to be "most perfect" by the Elizabethan playwright, Thomas Dekker. He included these lines in his 1604 play *The Honest Whore.*

> Of geometric figures the most rare,
> And perfect'st are the circle and the square,
> The city and the school much build upon
> These figures, for both love proportion.
> The city-cap is round, the scholar's square,
> To show that government and learning are
> The perfect'st limbs i'th' body of a state:
> For without them, all's disproportionate.
>
> (Dekker 272)

Spirals

Another important shape in mathematics is the spiral. In Chapter 2 we introduced the logarithmic spiral associated with the Golden Rectangle and found in the natural growth pattern of plants and shells. There are many types of spirals, and we will show just a few of them in Figure 3.7.

Name of Spiral	Equation	Shape	Some Examples
Logarithmic Spiral (Equiangular Spiral or Growth Spiral)	$r = ae^{b\theta}$	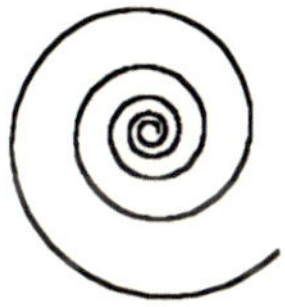	Nautilus Shell Snail Shell Sunflower Florets Elephant's Tusk Ram's Horn Rhinoceros' Horn Cat's Claw Pinecone Bracts Artichoke Bracts Pineapple Scales
Archimedes' Spiral	$r = a\theta$	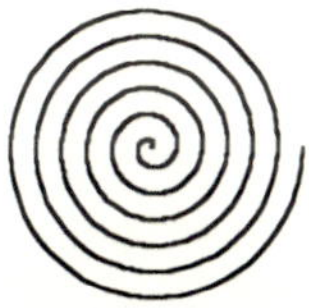	Coiled Rope Watch Spring Ionic Column Capitol Celtic Art Spirals
Hyperbolic Spiral (Inverse Spiral)	$r = \frac{a}{\theta}$	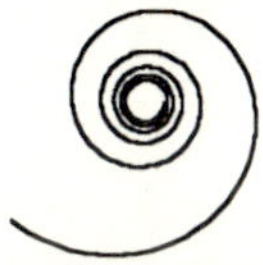	Fiddlehead Fern Eye of Horus Symbol
Fermat's Spiral (Parabolic Spiral)	$r^2 = a^2\theta$	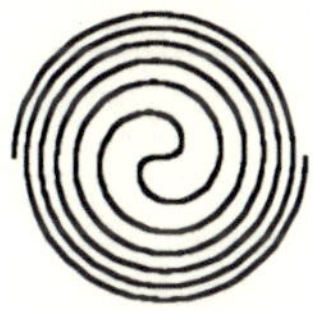	Celtic Art Double Spiral Mazes

Figure 3.7

Spirals are found throughout nature in the patterns of flowers, plants, and animals. These spirals of the natural world, shown in Figures 3.8 – 3.11, were photographed in the preserves and national parks of Costa Rica and Ecuador.

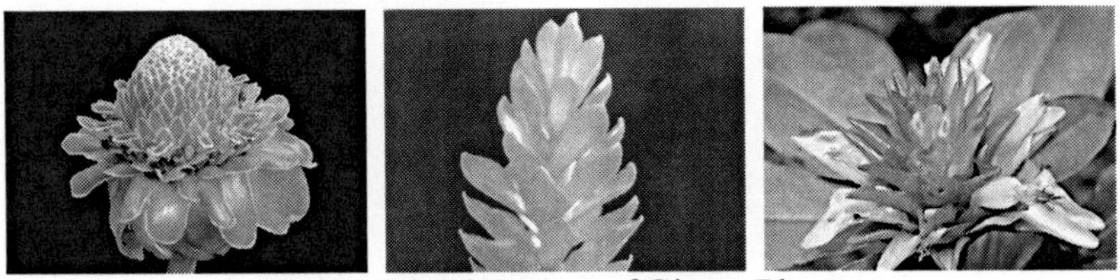

Spiral Growth Pattern of Ginger Plants

Figure 3.8

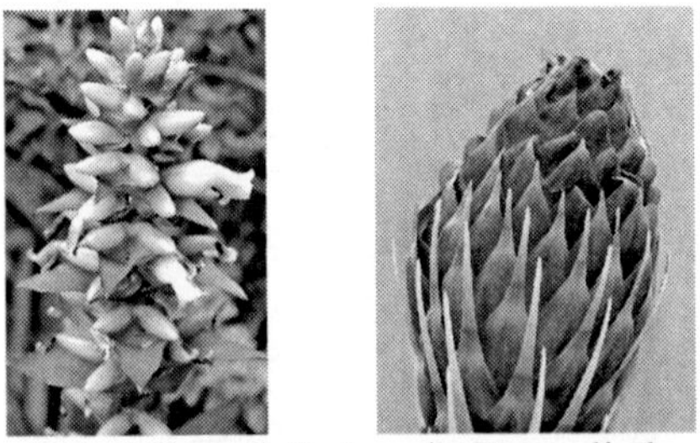

Spiral Growth Pattern in Bromeliads

Figure 3.9

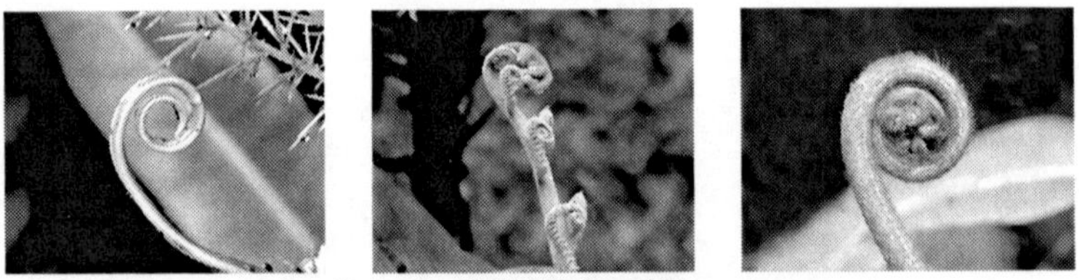

Spirals of Fiddlehead Ferns

Figure 3.10

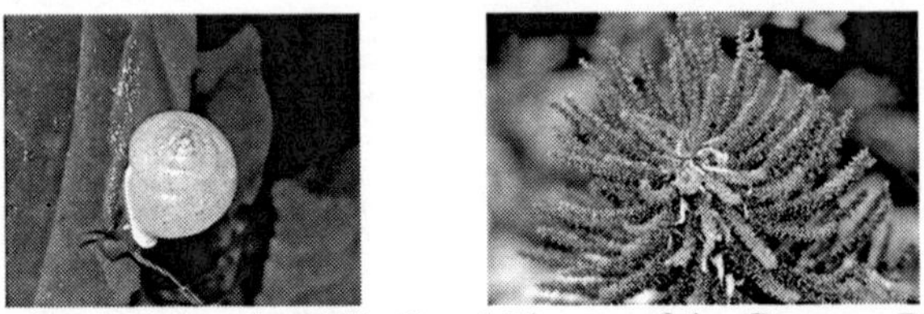

Spirals of Kinker Snail Shell and Flower of the Gunnera Plant

Figure 3.11

Spirals, mazes, and labyrinths are often the subject of poetry. Sometimes a poem is even written in such a shape. We turn to George Lynton Kaufman for a spiral poem that alludes to a medieval French verse form known as the *virelay*. In a virelay, the end rhyme of each stanza recurs as the first rhyme in the following stanza. Lynton uses repeating sound combined with the shape of a spiral to produce the form he calls "Spiralay."

SPIRALAY

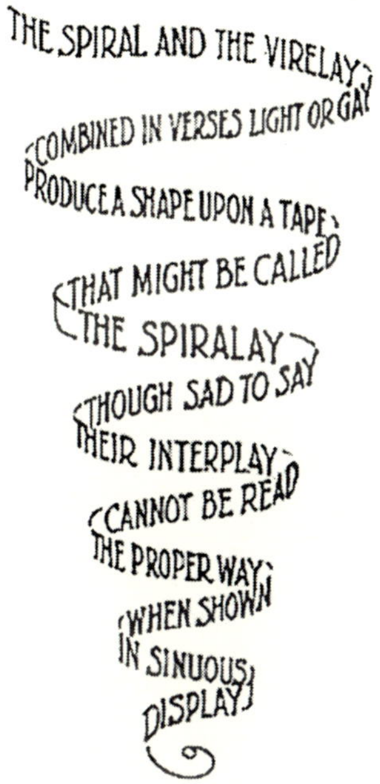

(Kaufman 41)

3.2 Shaped Poetry

Poetry literally takes on another dimension when poets consider the *visual* properties of the works they create. For centuries, there have been poets for whom the physical arrangement of letters and words is an integral part of the poem, whether the characters are hand-drawn, typed, or designed by a computer. There are many terms for this poetry, but in general, the terms "pattern poetry" or "shaped poetry" are used to refer to poems whose words are arranged in deliberate patterns. The impulse to arrange words in a way that is meaningful to the eye, as well as the ear, crosses time periods and cultures. There are some similarities in the pattern poems of different peoples, just as there are hundreds of one-of-a-kind pattern poems. And while some pattern poems may be read aloud without much difficulty, others — whether highly representational or abstract — cannot be read in a conventional manner but must be experienced as works of visual art. In the following pages, we will give a brief overview of pattern poetry, along with a few examples. Other pattern poems appear in illustrations throughout the book.

Labyrinths and Puzzles

In some pattern poems, the words are arranged to create a puzzle. For example, the poem may be an acrostic. That is, the first letters of each line spell out a word or message when the letters are read together. The SATOR word square, whose origins may go back to the 2nd century C.E., contains the five Latin words *sator*, *arepo*, *tenet*, *opera*, and *rotas* which may be read top to bottom, bottom to top, right to left, or left to right. It looks very much like the mathematical magic squares discussed in Chapter 1 and has been found in stone inscriptions throughout Europe.

S	A	T	O	R
A	R	E	P	O
T	E	N	E	T
O	P	E	R	A
R	O	T	A	S

Figure 3.12

A poem may also be written as a labyrinth, and the secret to the labyrinth must be understood before the reader can move inside. The following labyrinth by the French poet and musician Eustorg de Beaulieu is titled "Gloire à Dieu seul" ("Glory to God Alone"). It may be read from the center outward in all directions, following either a straight or zigzag pattern.

l u e s u e i d i e u s e u l
u e s u e i d a d i e u s e u
e s u e i d a e a d i e u s e
s u e i d a e r e a d i e u s
u e i d a e r i r e a d i e u
e i d a e r i o i r e a d i e
i d a e r i o l o i r e a d i
d a e r i o l G l o i r e a d
i d a e r i o l o i r e a d i
e i d a e r i o i r e a d i e
u e i d a e r i r e a d i e u
s u e i d a e r e a d i e u s
e s u e i d a e a d i e u s e
u e s u e i d a d i e u s e u
l u e s u e i d i e u s e u l

(de Beaulieu 85)

Religious belief, paradox, and mystery are often the subject of pattern poems. When these subjects are portrayed — or obscured — by the enigmatic properties of a word puzzle, readers may face quite a challenge in interpreting the poem.

Geometric Forms

Some pattern poems have also taken on familiar conventional forms, such as the cross, the altar, and the egg.[7] George Puttenham, a contemporary of William Shakespeare, provided Elizabethans with guidelines (shown in Figure 3.13) for using geometric shapes in poetry. In his 1589 work *The Arte of*

[7] These and other patterns have been documented by Dick Higgins, a collector and scholar of pattern poetry, who has studied examples from throughout the world and published the valuable compendium *Pattern Poetry: Guide to an Unknown Literature*. Albany, NY: SUNY Press, 1987.

English Poesie, Puttenham writes, "It is said by such as professe the Mathematicall sciences, that all things stand by proportion, and that without it nothing could stand to be good or beautiful" (64).

He goes on to say that "proportion Poeticall" rests in five qualities: staff (stanza), measure (meter), concord (rhyme), situation (placement of rhyme), and figure, or what Puttenham describes as the "ocular representation" (65).

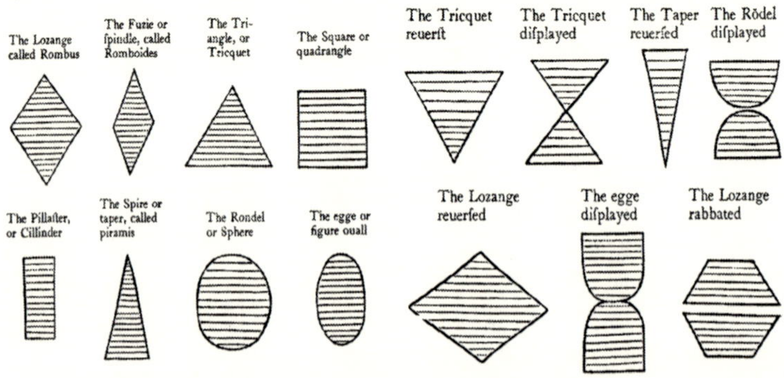

Figure 3.13

It's easy to see that several of the figures in Puttenham's chart (92) are variations or combinations of the regular plane figures, including triangles and quadrilaterals, that appear in the table of Figure 3.6. However, it is important to remember that Puttenham's diagram was not intended to be a guide to mathematicians, but a help for *poets*. Puttenham provides many illustrations of how the geometric forms may be used by poets, including this poem written in the form of a spire or taper:

Her Maieſtie, for many parts in her moſt noble and vertuous nature to be found, reſembled to the ſpire. Ye muſt begin beneath according to the nature of the deuice

Skie.

Azurd
in the
aſſurde,

And better,
And richer,
Much greter,

Crown & empir
After an hier
For to aſpire
Like flame of fire
In forme of ſpire

To mount on hie,
Con ti nu al ly
With trauel & teen
Moſt gratious queen
Ye haue made a vow
Shews vs plainly how
Not fained but true,
To euery mans vew,
Shining cleere in you
Of ſo bright an hewe,
Euen thus vertewe

Vanish out of our ſight
Till his fine top be quite
To Taper in the ayre
Endeuors ſoft and faire
By his kindly nature
Of tall comely ſtature
Like as this faire figure

(Puttenham 96)

Less than fifty years after Puttenham's work appeared, one of the most famous shaped poems in English was composed by George Herbert who arranged words to symbolize the poem's subject in "Easter Wings." Herbert originally printed the wing- shaped poem horizontally across the page, but it is often printed vertically, as shown below.

Easter-wings.

LOrd, who createdst man in wealth and store,
Though foolishly he lost the same,
Decaying more and more,
Till he became
Most poore:
With thee
O let me rise
As larks, harmoniously,
And sing this day thy victories:
Then shall the fall further the flight in me.

My tender age in sorrow did beginne:
And still with sicknesses and shame
Thou didst so punish sinne,
That I became
Most thinne.
With thee
Let me combine
And feel this day thy victorie:
For, if I imp my wing on thine,
Affliction shall advance the flight in me.

(Herbert 43)

Modern Pattern Poetry

In modern times, the creators of pattern poetry have concentrated on using words as a *medium*, drawing on their abstract qualities, their appearance, and their literal and suggested meanings. Modern poets have gone beyond the practice of simply arranging words to create puzzles, geometric patterns, or symbolic shapes. The following sampler of poetry composed since the beginning of the twentieth century reveals a complex intermingling of poetry and the visual arts.

The French poet Guillaume Apollinaire used the term "calligrammes" for poems in which he arranged words to create a visual image of the poem's subject. In the poem "Il Pleut" ("It's Raining") the downward-flow of letters intensifies the image of falling rain drops described in the poem. This calligramme also uses visual properties in ways that subtly draw us into the ideas of the poem. By arranging the letters vertically, Apollinaire forces the reader to slow down and read with care, thus making the process of reading more deliberate and self-conscious, just as the rain makes the speaker pause and reflect on voices and encounters from his past.

IL PLEUT

Il pleut des voix de femmes comme si elles étaient mortes même dans le souvenir
c'est vous aussi qu'il pleut merveilleuses rencontres de ma vie ô gouttelettes
et ces nuages cabrés se prennent à hennir tout un univers de villes auriculaires
écoute s'il pleut tandis que le regret et le dédain pleurent une ancienne musique
écoute tomber les liens qui te retiennent en haut et en bas

(Apollinaire 171)

This English translation of "Il Pleut" was done by Roger Shattuck:

> It's raining women's voices as if they had died even in memory
> And it's raining you as well marvelous encounters of my life
> O little drops
> Those rearing clouds begin to neigh a whole universe of auricular cities
> Listen if it rains while regret and disdain weep to an ancient music
> Listen to the bonds fall off which hold you above and below
> (Shattuck 170)

The Swiss poet Eugen Gomringer called his first works of "concrete poetry" *constellations*, in which he brought together clusters of words and graphic elements in "response to a particular creative impulse" (Solt 9).

o
bo
blow
blow blow
blow blow blow
blow blow
blow
bo
o o
go so
grow show
grow grow show show
grow grow grow o show show show
grow grow show show
grow show
go so
o o
lo
flow
flow flow
flow flow flow
flow flow
flow
lo
o

(Gomringer 94)

The American concrete poet Mary Ellen Solt combines the letters for the word "FORSYTHIA" with their Morse code symbols in an acrostic-based poem, creating an exuberant image of the forsythia plant. The poem was originally printed on an intense yellow background to intensify its visual impact.

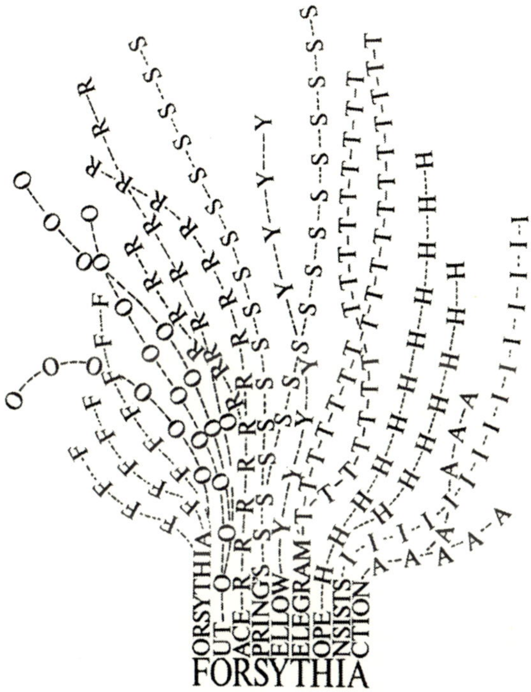

(Solt 243)

In some modern pattern poetry, the use of shaped language results in visual wordplay, as in Dorthi Charles' "Concrete Cat" where the whimsical arrangement of letters not only portrays the cat but also tells a story.

Concrete Cat

A A
e r e r

eYe eYe stripestripestripestripe t
whisker whisker stripestripestripe a
whisker m h whisker stripestripestripestripes i l t a i l
o t stripestripestripe
U stripestripestripestripe

paw paw paw paw mouse

dishdish litterbox
litterbox

(Charles 606)

Long after George Puttenham's guide for poets appeared, we still see pattern poems whose words are shaped in imitation of ideal forms, as in Denis Garrison's poem "Nautilus." Here the shell's properties are captured in both language and form: Garrison simultaneously writes *about* the nautilus while arranging his words in a pattern that suggests the *shape* of the nautilus. You may notice that the poet also experiments with a now-familiar number sequence. By using an increasing number of syllables in each succeeding line, beginning with the first line, Garrison creates a Fibonacci sequence in the first seven lines, where the numbers of syllables, are, respectively, 1-1-2-3-5-8-13. In the last seven lines of the poem, we can observe a descending Fibonacci sequence, as the number of syllables decreases: 13-8-5-3-2-1-1.

One
time
in its
pulsing span,
the nautilus must
rise from deep reefs to glimpse the sun
gold-shimmer on the bright silver ceiling of the sea.

It cannot long abide in dancing columns of blue light
but must find its home in the lightless realm of dayless night.
There, in endless reaches of black sea, it preys ceaselessly.
Time does not pass but it floats by imperceptibly
until, finally, the failed molluskan pulse takes flight.

Then, up from the inky fastness of its coral caves,
robed in many-chambered splendor,
glorious in death,
nautilus
ascends
once
more.

(Garrison)

In other chapters, you will find more examples of pattern poetry, including Mary Ellen Solt's "Moon shot" (Chapter 2) and Rodrigo Siqueira's "The Cantor Dust" (Chapter 4), where contemporary writers have used visual *symbols* as well as words to capture ideas as enigmatic as those contained in early labyrinthine poems.

3.3 Symmetry in Mathematics

The repeated refrain of a song, the reflection of a mountain in a lake, the Islamic tiling patterns of the Alhambra in Spain, the columns of the Parthenon in Greece, the five petals on a wildflower — all are examples of symmetry. Symmetry is a fundamental organizing principle of shape. It helps in classifying and understanding patterns in mathematics, nature, art, and, of course, poetry. And often the counterpoint to symmetry — the breaking or interruption of symmetry — is just as important in creative endeavors.

Symmetry can be defined in many ways, depending upon the context of language or discipline. When we hear the word "symmetry," we often think of reflections or mirror images, such as those seen below in the wings of a Zebra butterfly and a Saturn moth.

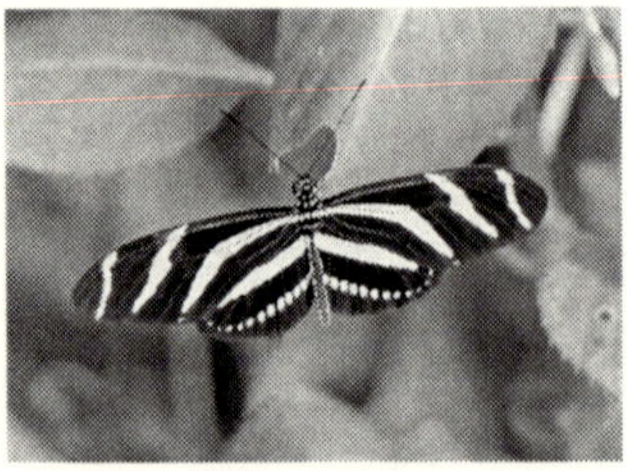

Figure 3.14

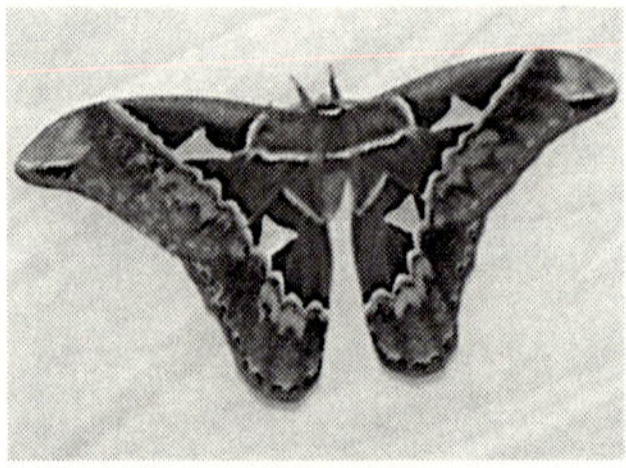

Figure 3.15

In a more general context, symmetry refers to similarity, balance, regularity or proportion of form. We might refer to the symmetry of ideas in the stanzas of a poem, the symmetry of space created by the placement of furniture in a room, or the proportion in the architecture of a skyscraper. Historians refer to the symmetry of global involvement during the two World Wars, while cognitive scientists write about the symmetry of analogous ideas. Poets have employed symmetry in the shape, rhyme scheme, repetition, and ideas of their works.

Basic Symmetries

In contrast to the broadly interpreted ideas of balance and proportion, mathematicians have a very precise definition of symmetry. Mathematically speaking, an object (or set of points in the plane) is symmetric if it remains unchanged or *invariant* under a certain types transformations or movements in the plane. These movements include rotation, reflection, and translation along a straight line and are illustrated below in Figures 3.16 — 3.19. After transformation, the set of points representing the new object must be identical to the set of points representing the original object. To illustrate this, we see that the letter E is not symmetric when reflected about a vertical line down its middle, but it is symmetric when reflected about a horizontal line across its middle, as shown in Figure 3.16.

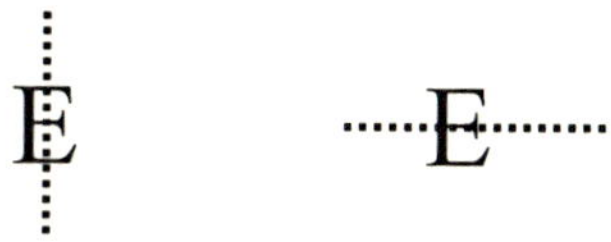

Figure 3.16

A circle is symmetric under rotation about its center, since any such rotation leaves it looking exactly the same. The circle shown in figure 3.17 has an arrow drawn on one radius so that the rotations may be visualized. The mirror image of the circle reflected across any diameter also leaves its shape unchanged.

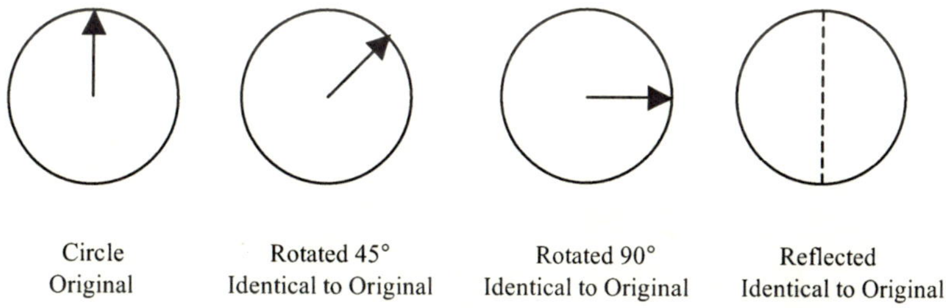

Figure 3.17

A square is somewhat different in its symmetry, since only certain rotations and reflections will leave it invariant, as seen in Figure 3.18. When a square is rotated 45°, the resulting figure not only looks different from the original, but the set of points in the plane representing the diamond shape is not the same set that represented the original square. Only when the original square is rotated 90°, 180°, 270°, or any other integer multiple of 90°, does the resulting shape look the same and consist of the same set of points in the plane. Similarly, only certain reflections leave the square invariant.

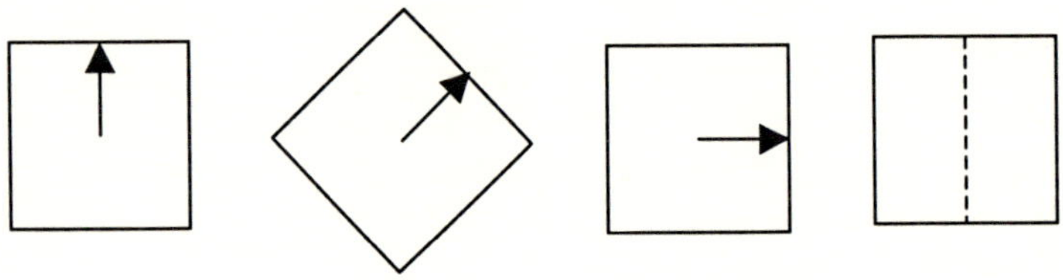

Square Original	Rotated 45° Different from Original	Rotated 90° Identical to Original	Reflected Identical to Original

Figure 3.18

Mathematicians describe four basic symmetries of the plane that can leave the size and shape of an object unchanged: translation, rotation, reflection, and glide reflection. This last symmetry is really a combination of translation and reflection. The technical term for one of these basic symmetries is an *isometry* (from the Greek *isos*, meaning equal and *metron*, meaning measure). A visual presentation of the four basic isometries is found in Figure 3.19.

Type of Symmetry	Symmetry Illustration	Explanation
Translation	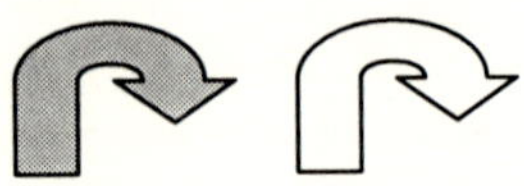	The original shape (gray) is copied (white) and moved along a straight line from its original position. The shape is not turned in any way from its original orientation.
Rotation	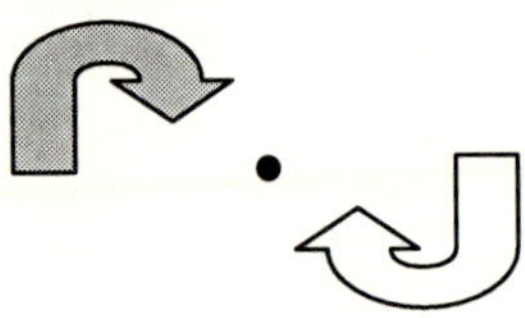	The original shape is rotated about a point. In this example the original shape is rotated 180 degrees about the black circle.

Reflection	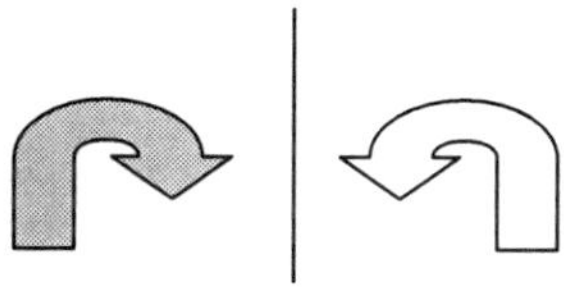	The original shape is reflected across the "mirror" of the straight line.
Glide Reflection	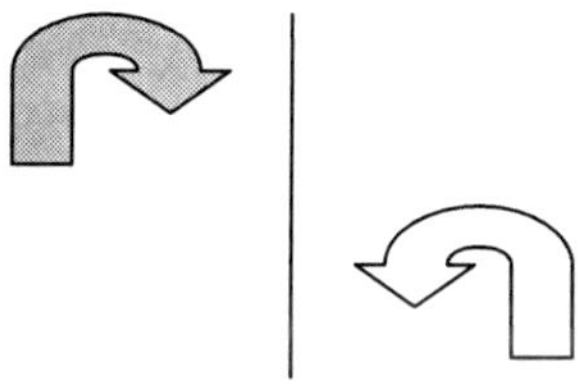	The original shape is first reflected across a "mirror" line and then translated to its final position.

Figure 3.19

Translation symmetries are commonly found in architectural details such as friezes, and in weaving and quilting patterns, tile patterns, and wallpaper patterns, especially borders. Translations also appear in nature, as seen in the branches of a Costa Rican fishtail palm (Figure 3.20) and trees in a grove of Colorado aspens (Figure 3.21).

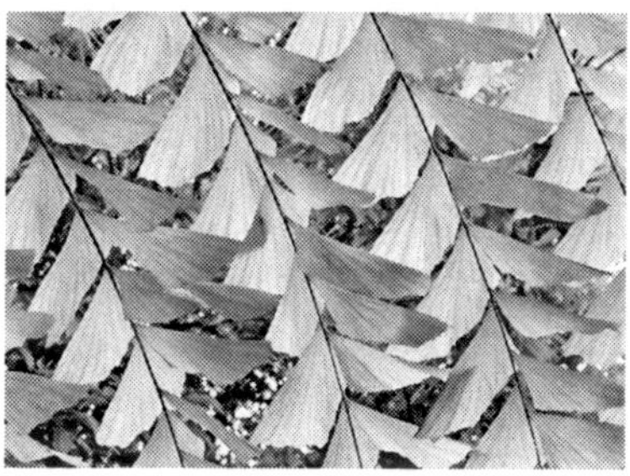

Figure 3.20

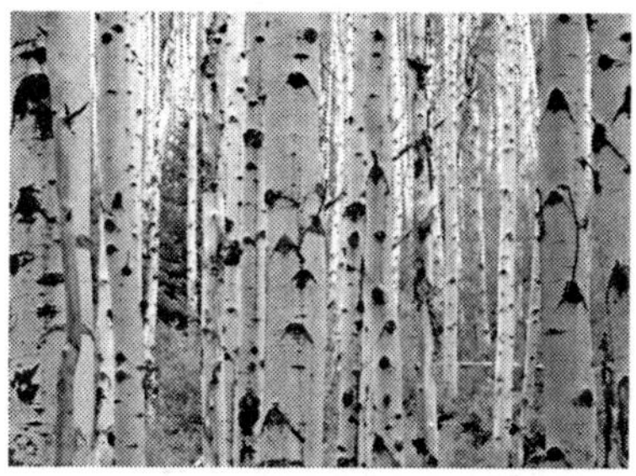

Figure 3.21

The images below of a nest of pileated woodpecker babies provide some fun examples of translational symmetry as the fledglings eagerly await a parent to feed them. Of course, the tiny birds are only approximating translational symmetry, but it is interesting to see how they move their heads in tandem as they search for and call to their parents.

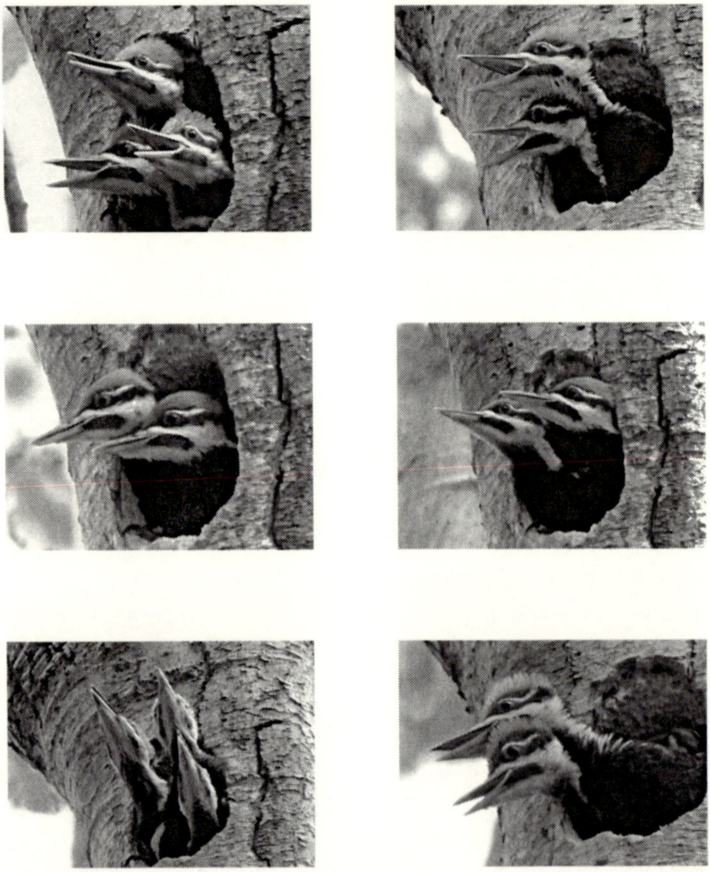

Figures 3.22

There are many types of rotational symmetry, and they are usually named for the number of repetitions of the original shape. Since a circle has 360 degrees, mathematicians refer to *n-fold rotational symmetry*, where *n* divides 360 into equal parts. A 6-fold rotational symmetry would divide a circle into 6 equal parts, each of 60 degrees, and would exhibit 6 rotations of the same shape. Similarly, a 20-fold rotation would divide a circle into 20 equal parts, each of 18 degrees and would exhibit 20 rotations of the same shape. Some examples of rotational symmetries are shown in Figure 3.23.

Type of Rotational Symmetry	Shape is Rotated	Instances of the Shape	Example
2-fold	180°	2	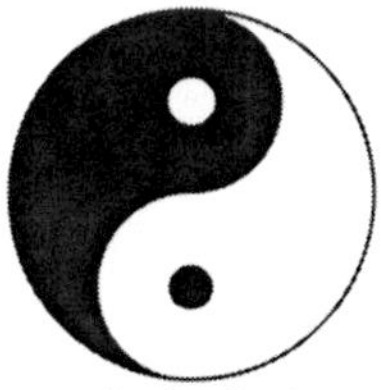 Chinese symbol of Yin and Yang
3-fold	120°	3	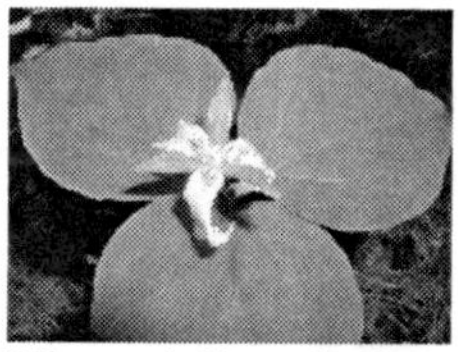 Trillium, Adirondack Park
4-fold	90°	4	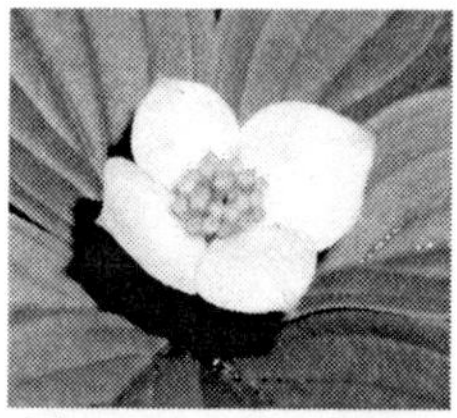 Bunchberry, Adirondack Park
5-fold	72°	5	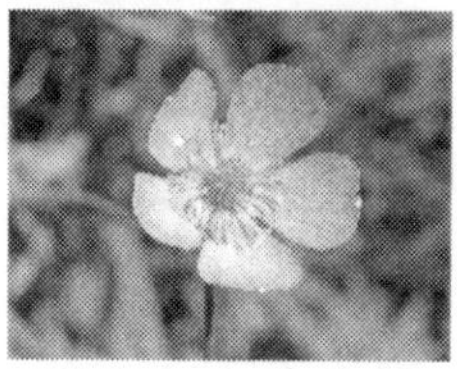 Buttercup, Ireland

6-fold	60°	6	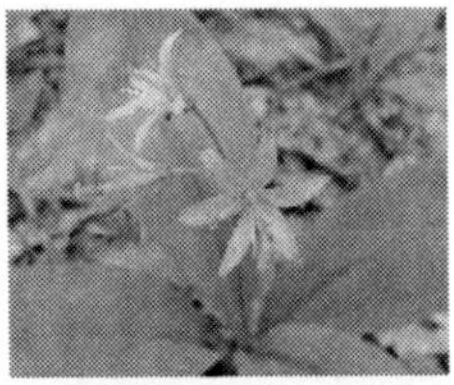 Bluebead Lily, Adirondack Park
15-fold	24°	15	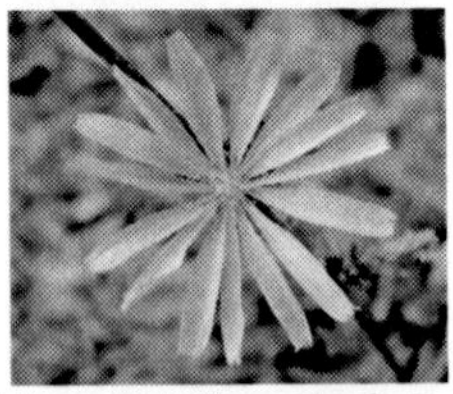 Chicory, Adirondack Park
20-fold	18°	20	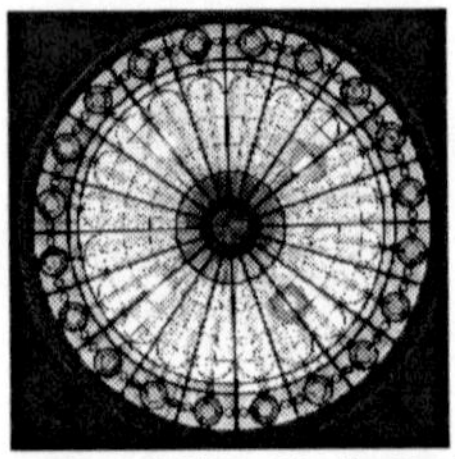 Rotunda of City Hall, Dublin

Figure 3.23

Reflection symmetries (often called bilateral symmetries) are quite common in nature, as seen in the mirror image of a mountain along the coast of Iceland (Figure 3.24) or the sidewall of Santa Elena Canyon echoed in the Rio Grande River at Big Bend National Park (Figure 3.25).

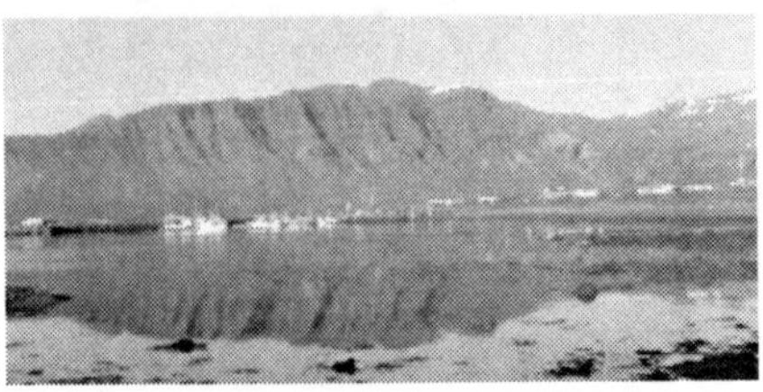

Figure 3.24

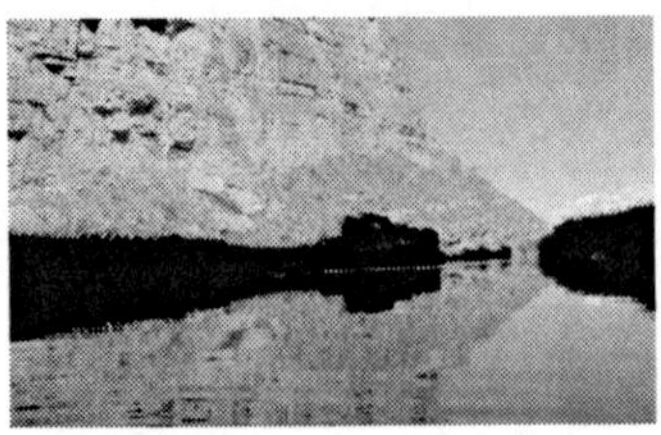

Figure 3.25

More bilateral symmetry is seen in the echo of wing patterns of a swallowtail butterfly from Adirondack State Park, New York (Fig. 3.26) and a canopy of tree ferns in the Monteverde Cloud Forest of Costa Rica (Fig. 3.27).

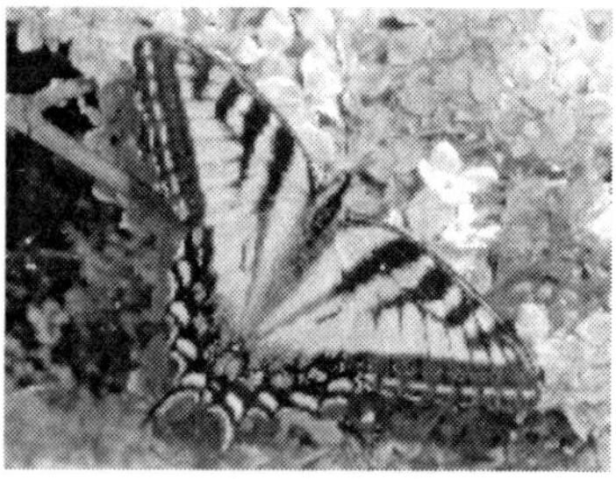

Figure 3.26

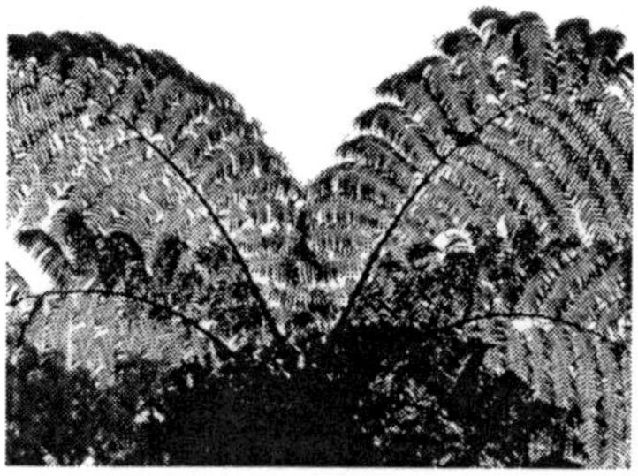

Figure 3.27

Architects often use bilateral symmetry in their designs, as seen below in these two modern churches of Iceland, the one on the left in Reykjavik (Figure 3.28) and the one on the right in Akureyri (Figure 3.29). The pleasing balance that reflections bring to these buildings are not limited to modern architecture. The harmony of reflection on either side of a central axis is found in examples that cross cultures and time periods, including the Greek Parthenon built in the 5th century B.C.E., the Renaissance churches of Europe built in the 15th century, and the palaces and Buddhist pagodas built throughout Chinese history.

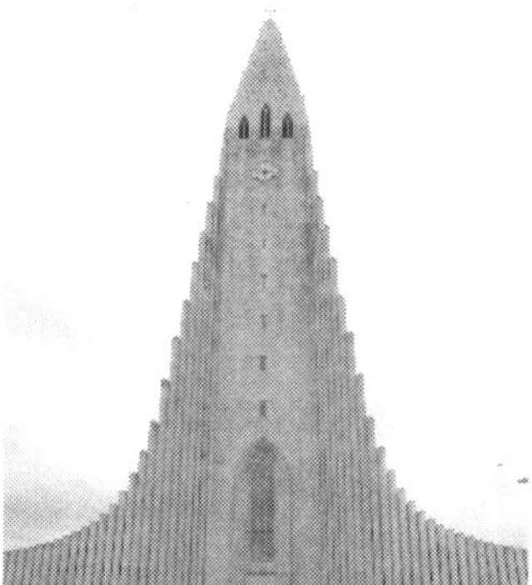

Figure 3.28

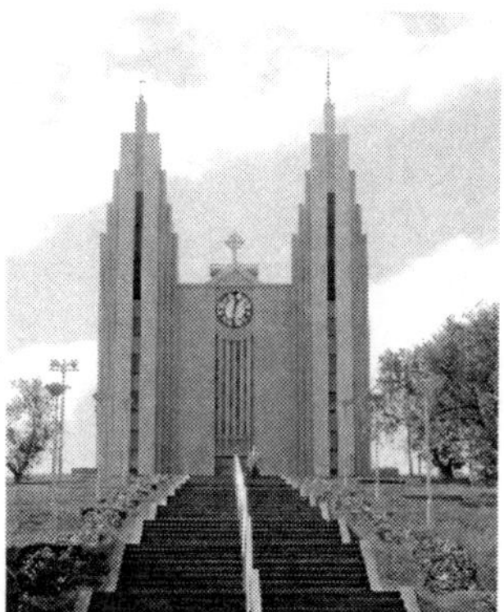

Figure 3.29

Finally, glide reflection symmetry is often found in plants, such as those in Figures 3.30 and 3.31, photographed in Costa Rica. On the left we see heliconia, and on the right is the epiphytic growth of leaves on a tree trunk in the rainforest.

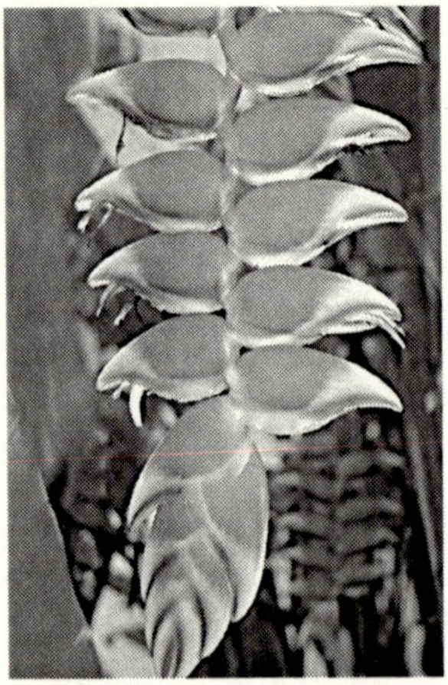

Figure 3.30

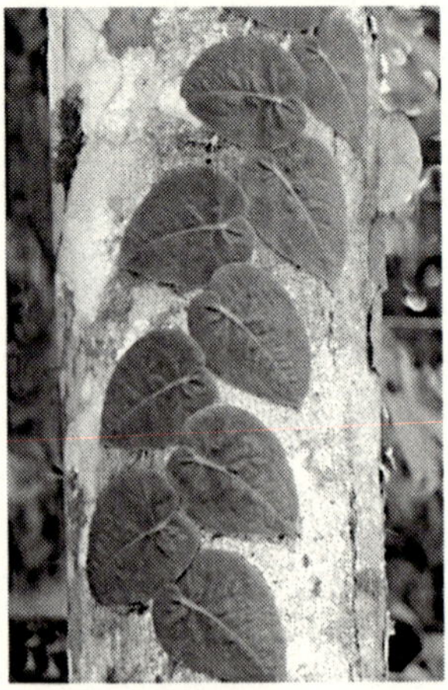

Figure 3.31

Not only do mathematicians classify these symmetries or isometries into the four basic types (translation, rotation, reflection, and glide reflection), but they also categorize the ways in which these symmetries can be combined along a line and in the plane. Line symmetries, often called frieze patterns, are symmetric designs in one direction that form a linear pattern or border. There are only 7 different ways to combine translation, rotation, reflection, and glide reflection in a frieze, as shown in Figure 3.32.

Translation	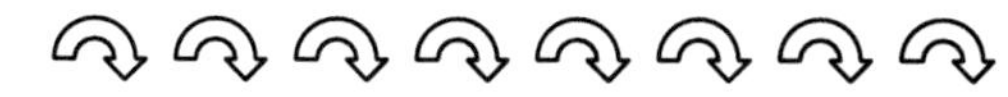
Glide reflection	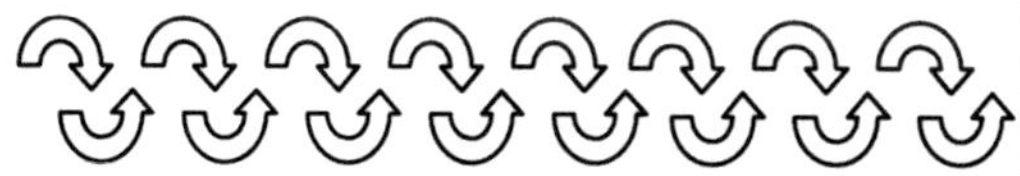
Reflections that are translated (many reflection lines)	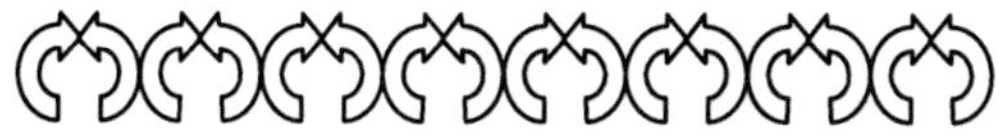
Translations that are reflected (only one reflection line)	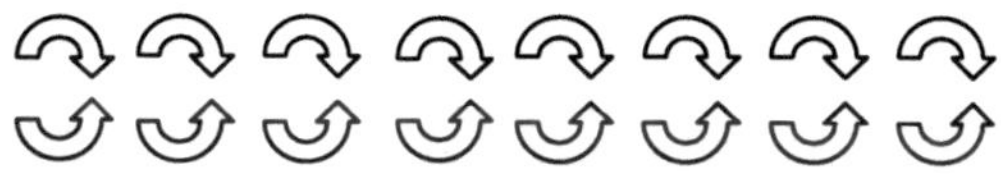
Translation of 2-fold rotation	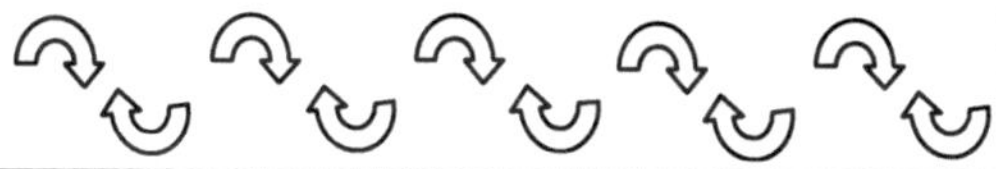
Reflection of a 2-fold rotation	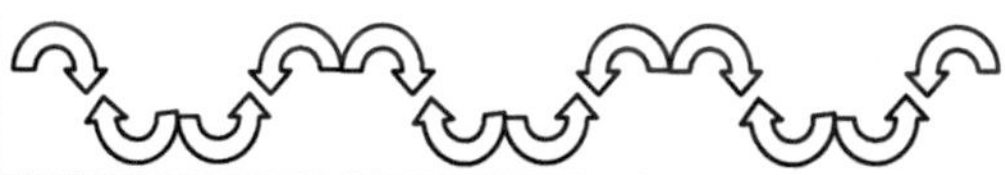
Reflections in both directions (sets of translated 2-fold rotations)	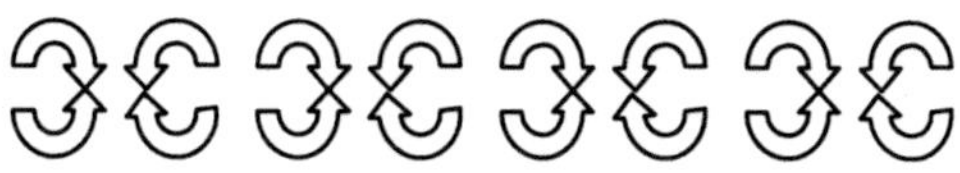

Figure 3.32

If we tried to classify all the ways in which the plane can be covered in two directions by the four basic isometries, there would be only 17 possible arrangements. Three of these coverings of the plane, known as wallpaper groups, are shown below in Figure 3.33.

Simple translational symmetry in two directions	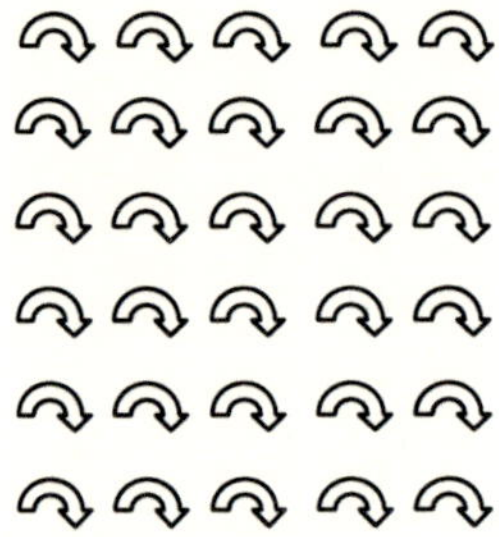
Glide reflection symmetry with translations	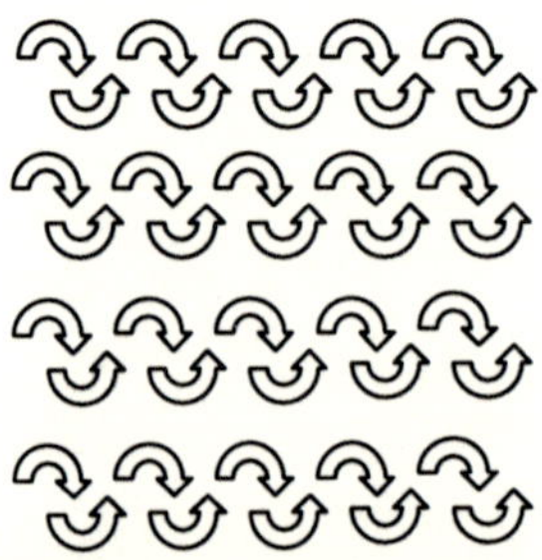
Two-fold rotations, reflections, and translations	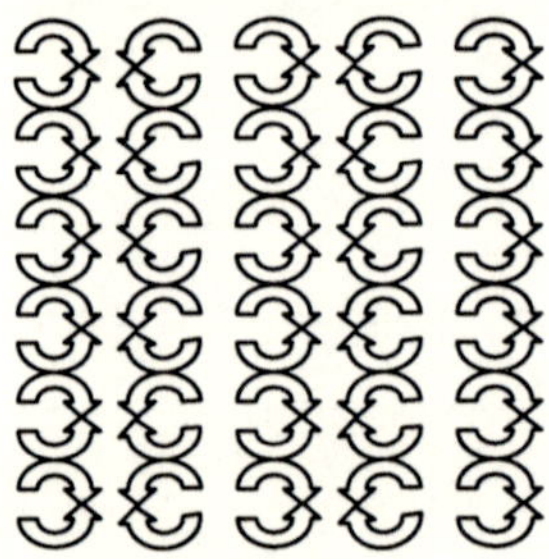

Figure 3.33

Visual symmetry is pleasing to the eye. It feels balanced and complete, helping our intuition fill in missing details. Sometimes, however, perfect symmetry may feel static or repetitious. Creative manipulation of fixed forms, such as those that are symmetric, can result in new ways of seeing, describing and thinking. Let's see how that happens with tessellations, compositions that rely heavily upon symmetry.

Tessellations and the Art of M. C. Escher

A *tessellation* is an arrangement of congruent shapes completely filling a plane surface (for example, a piece of paper). If the shapes that tessellate the plane are regular polygons, such as triangles, squares, or hexagons, the tessellation is called a *regular tessellation.*

Regular Tessellation of Triangles	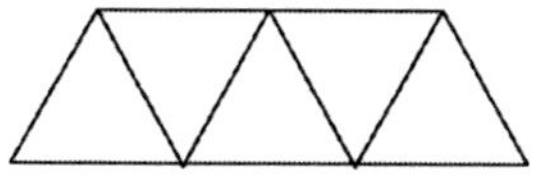
Regular Tessellation of Squares	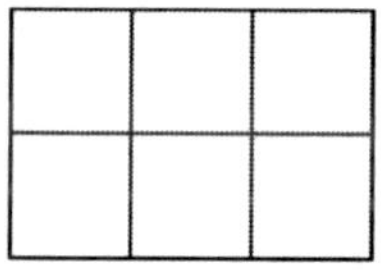
Regular Tessellation of Hexagons	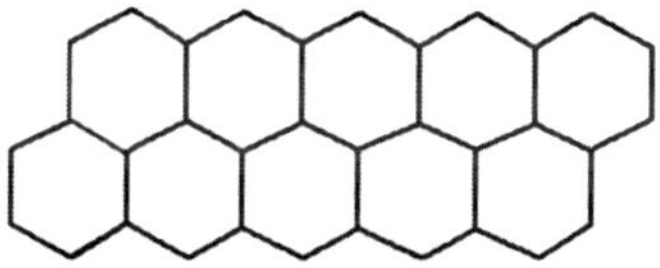
Irregular Tessellation of Hexagons	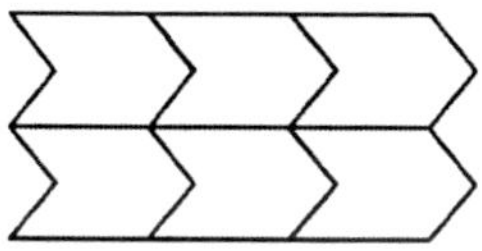
Tessellation of One Irregular Shape	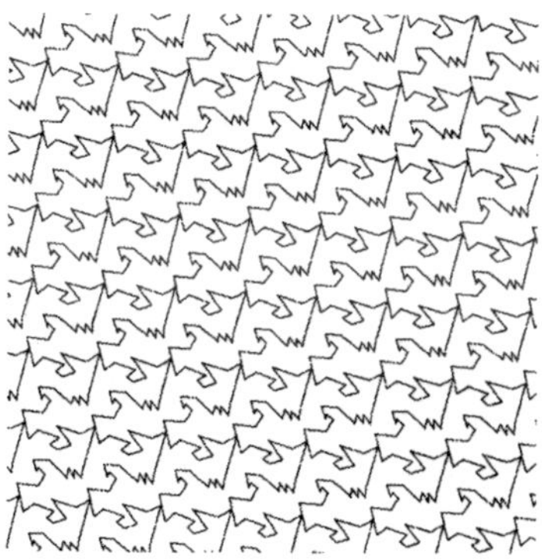

Tessellation of a Combination of Irregular Shapes	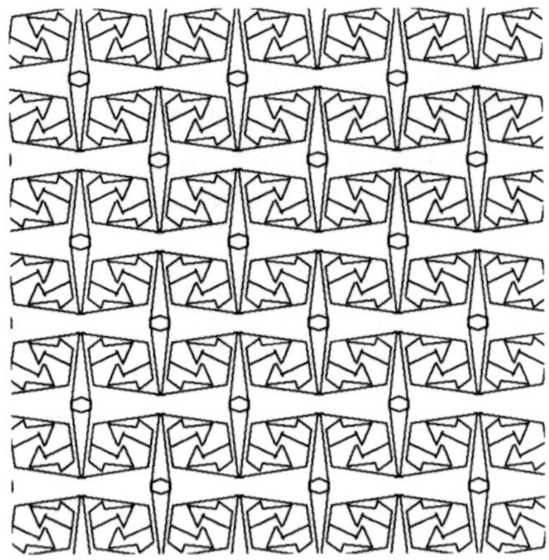

Figure 3.34

The most famous tessellations were created by the Dutch graphic artist, M. C. Escher. Although he had no mathematical training, his work employed many mathematical ideas, developed in his own notes, without a formal knowledge of geometry or symmetry. His works have been intensely studied by mathematicians and are found in the illustrations of countless math books. Escher used complex figures or combinations of figures to tile the plane, creating sophisticated patterns of symmetry. Escher used color, shading, scale, irony, and even whimsy in his creations. He wondered about his fixation on experimenting with and creating new and different types of tessellations – and more importantly he asked himself, “Why does none of my fellow-artists seem to be fascinated as I am by these interlocking shapes?” (MacGillavry vii). Escher wrote of his pleasure in creating these increasingly intricate jigsaw puzzles and described how the fanciful creatures in his tessellations themselves seemed to inspire and even control his composition. In the course of creating over one hundred and fifty tessellations, Escher explored balance and imbalance, symmetry and asymmetry, contrasts of color, and black and white duality. Figure 3.35 illustrates many of these concepts.

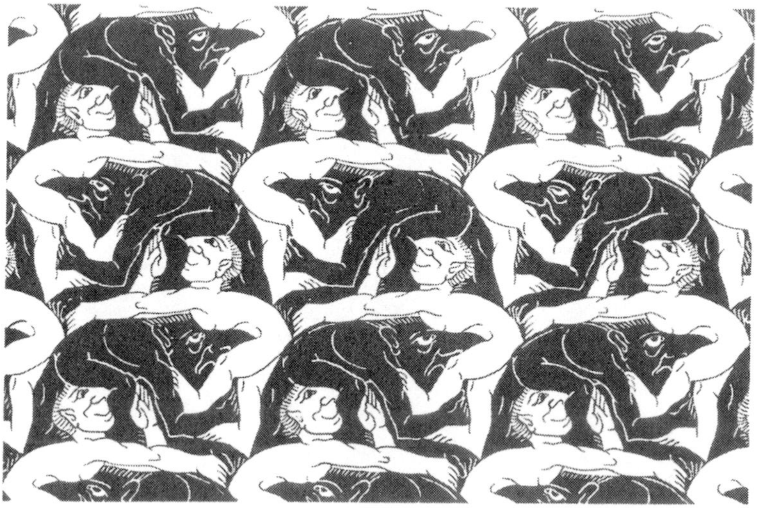

M. C. Escher's "Symmetry Drawing E63" © 2007 The M.C. Escher Company - Holland. All Rights reserved. www.mcescher.com

Figure 3.35

Escher's work reached a new level of creativity when he began drawing his curious creatures on a progressively smaller scale. He met the mathematician H. S. M. Coxeter and asked for an explanation of how Coxeter was able to make drawings on a circle that are similar in shape, but grow smaller as they approach the circumference of a circle. Escher never fully understood the hyperbolic geometry Coxeter was using, but he was able to extrapolate from his previous works that tiled the Euclidean plane to tiling a hyperbolic disc. Without mathematical background, but with an eye that saw how to curve and shrink his figures, Escher created many examples of tessellations in hyperbolic space. One example, "Circle Limit III," is shown below in Figure 3.36.

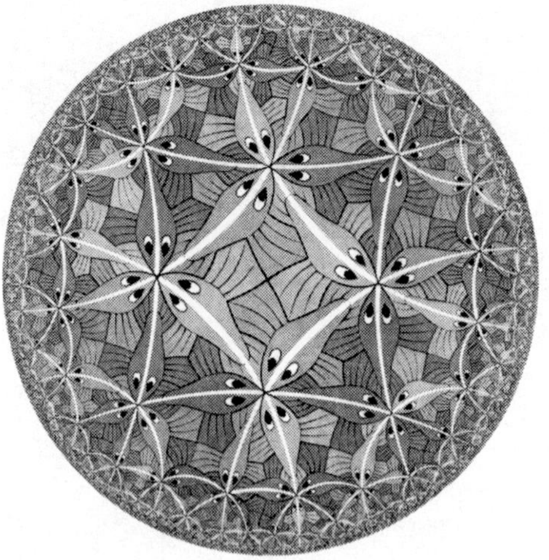

Figure 3.36

In order to explain Escher's later tessellations, we will present a brief introduction to hyperbolic geometry, starting from the familiar Euclidean geometry we learned in school. In Euclidean geometry, if we have a line L and a fixed point P that is not on L, there is one and only one line parallel to L that passes through the point P.

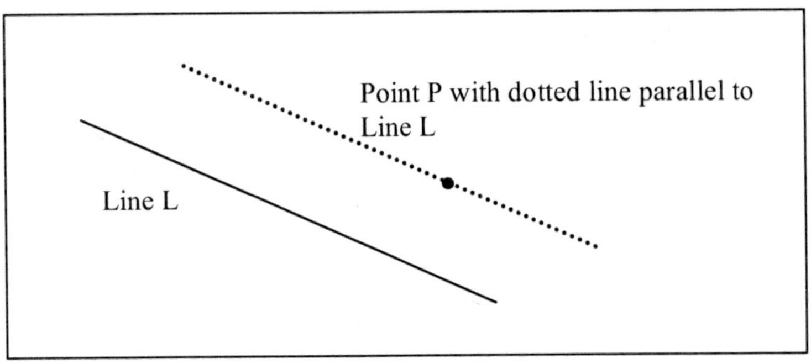

Figure 3.37

On the other hand, in a hyperbolic plane there are an infinite number of lines passing through a fixed point P that are parallel to the given line L. The terms *line* and *plane* are used differently in hyperbolic geometry. The *plane* is a disc or circle, and *lines* are arcs that intersect the circumference of the circle at right angles. *Parallel lines* in this geometry are non-intersecting arcs. In Figure 3.38, lines AB and CD are parallel, since they are not intersecting. Line PQ is not parallel to AB, but it is parallel to line CD.

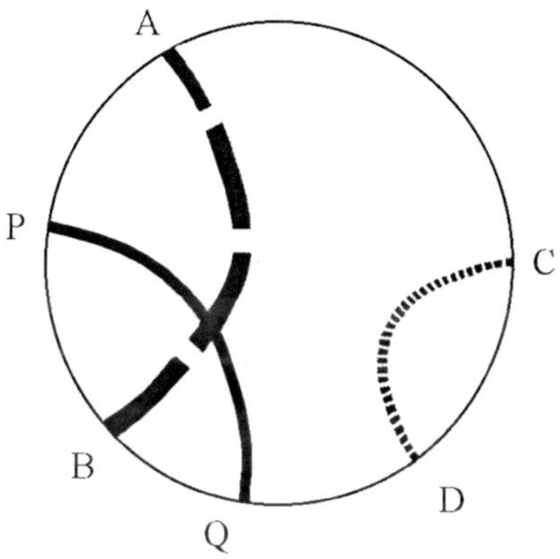

Figure 3.38

On the hyperbolic disc, it appears that the lines stop at the edge of the disc, but they really continue infinitely. As images get closer to the edge of the disc, they will gradually appear smaller as they bend to fit on the disc, but in the hyperbolic world they are actually the same size. This is illustrated in Figure 3.39, where hyperbolic rectangles are drawn on the hyperbolic disc. Although the central rectangle is symmetric, as we move toward the edges of the disc, the rectangles distort and grow smaller.

Hyperbolic geometry

plane = circle
line = arc at rt. angle to circum.
parallel line = non intersecting arc

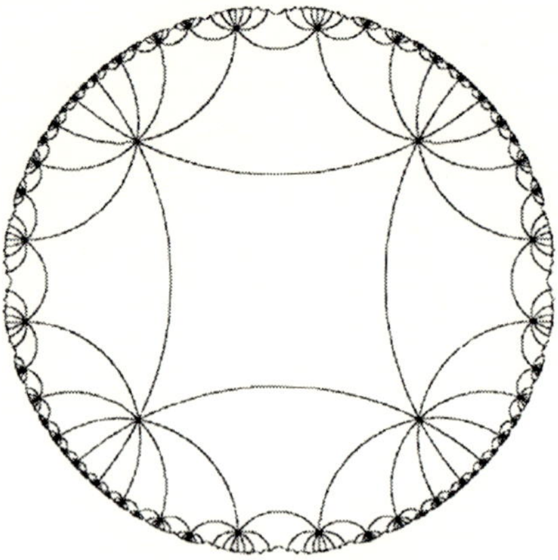

Figure 3.39

The genius of Escher is evident in both the beauty and mathematical complexity of his works. His ability to create intricate, symmetric tessellations (as seen in Figure 3.35), as well as to manipulate that symmetry into a new form (as seen in Figure 3.36), crossed disciplinary boundaries with both artistic and geometric innovation. Countless math texts, art books, and works in other fields use Escher's famous images for illustration and explanation. In the foreword of *Escher: Visions of Symmetry,* Douglas Hofstadter uses the word "poetry" to characterize Escher's tessellations. He explains Escher's visual poetry by saying,

> Who can say what the fount of Escher's inspiration truly was? I can make but a crude stab at it. It seems rooted in his joy in discovering the subtle in the mundane and in his unshakable belief that pattern is sublime. It springs from his passionate pursuit of the visual poetry around him, and his irrepressible urge to share this poetry with others by passing it through the filter of his eye, his hand, his mind, and his heart (Schattschneider viii).

3.4 Symmetry in Poetry

Although it may not at first be as visually apparent as the symmetries in Escher's work, symmetry plays a vital role in poetry, too. We can find many instances of symmetry in the physical shape of a poem. We can also find symmetry in the way a poem's rhyme scheme or language repeats. Poets have made use of the visual, tactile, and logical elements associated with symmetry; they have created and then purposely broken or interrupted their own symmetrical patterns; and they have explored the metaphorical possibilities of the *idea* of symmetry.

Symmetry in Structure

In "Sence You Went Away," James Weldon Johnson, lawyer, novelist, and poet of the Harlem Renaissance, relies on symmetry to develop the poem's mournful tone.

Seems lak to me de stars don't shine so bright, a
Seems lak to me de sun done loss his light, a
Seems lak to me der's nothin' goin' right, a
Sence you went away. b

Seems lak to me de sky ain't half so blue, c
Seems lak to me dat ev'ything wants you, c
Seems lak to me I don't know what to do, c
Sence you went away. b

Seems lak to me dat ev'ything is wrong, e
Seems lak to me de day's jes twice es long, e
Seems lak to me de bird's forgot his song, e
Sence you went away. b

Seems lak to me I jes can't he'p but sigh, g
Seems lak to me ma th'oat, keeps gittin' dry, g
Seems lak to me a tear stays in ma eye, g
Sence you went away. b

(Johnson 41)

Johnson creates a repetitive structure that emphasizes how the speaker's grief continues through time. Even on first reading, we can observe what a mathematician might call translational symmetry in the physical shape of this poem. That is, the original shape (the first stanza) is repeated three times in a straight line down the left margin. There are other symmetries as well: within each stanza, the rhyme scheme is similar (a,a,a,b/c,c,c,b/d,d,d,b, etc.), and the first four words and the last line of each stanza are identical.

Despite the symmetries in the poem's structure, we find interruptions in the symmetry when we examine its language. With the exception of the phrase "I don't know what to do," the first three stanzas speak about changes to the natural world "sence you went away." The stars, the sun, the sky, and the bird have all lost their vibrance, and the day is twice as long. The impact of the speaker's loss on his world is summed up in the lines, "'der's nothin' goin' right...' and 'ev'ything is wrong....'"

In the last stanza, the poem's focus shifts from the domain of the natural world to the personal. Breaking the pattern that has been set up, the speaker no longer describes the departure of his loved one by showing how it is reflected in the natural world; instead, he reveals his own response, concentrating on the physical manifestations of his dry throat and unstoppable tears. While the symmetry of the first three stanzas builds and enforces the melancholy tone of this poem, the interruption of that symmetry makes the tone more poignant, more personal. The strength of the poem lies in its simplicity and in the way it sets up and then breaks a symmetrical pattern. By the fourth stanza, we know the structure, but what is new to us — and not revealed until the end — is the speaker's own response to his loss.

In "Small Hands, Relinquish All," Edna St. Vincent Millay also uses broken or interrupted symmetry with powerful results.

Small hands, relinquish all:
Nothing the fist can hold,—
Not power, not love, not gold—
But suffers from the cold,
And is about to fall.

The mind, at length bereft
Of thinking, and its pain,
Will soon disperse again,
And nothing will remain:
No, not a thought be left.

Exhort the closing eye,
Urge the resisting ear,
To say, "The thrush is here";

To say, "His song is clear";
To live, before it die.

Small hands, relinquish all:
Nothing the fist can hold,
Not power, not love, not gold,
But suffers from the cold,
And is about to fall.

The mind, at length bereft
Of thinking and its pain,
Will soon disperse again,
And nothing will remain:
No, not a thing be left.

Only the ardent eye,
Only the listening ear
Can say, "The thrush was here!"
Can say, "His song was clear!"
Can live, before it die.

(*Collected Poems*, 441-42)

At first glance, the poem exhibits a structural symmetry somewhat similar to that in James Weldon Johnson's poem. The rhyme scheme of each stanza is abbba, cdddc, etc., and the first and fifth lines of each stanza rhyme. When we look more closely, the second set of three stanzas appears to be a repetition of the first set. Is the second half of the poem identical to the first, thus perhaps creating a kind of translational symmetry? We can only answer the question when we place the two halves of the poem side by side. Then, subtle differences become apparent:

(Stanza 1)	(Stanza 4)
Small hands, relinquish all:	Small hands, relinquish all:
Nothing the fist can hold,—	Nothing the fist can hold,
Not power, not love, not gold—	Not power, not love, not gold,
But suffers from the cold,	But suffers from the cold,
And is about to fall.	And is about to fall.
(Stanza 2)	(Stanza 5)
The mind, at length bereft	The mind, at length bereft
Of thinking, and its pain,	Of thinking and its pain,
Will soon disperse again,	Will soon disperse again,

And nothing will remain: No, not a thought be left.	And nothing will remain: No, not a thing be left.
(Stanza 3) Exhort the closing eye, Urge the resisting ear, To say, "The thrush is here"; To say, "His song is clear"; To live, before it die.	(Stanza 6) Only the ardent eye, Only the listening ear Can say, "The thrush was here!" Can say, "His song was clear!" Can live, before it die.

Small changes begin to appear as soon as we compare stanzas 1 and 4. The significance of the changes increases slowly, and their impact on the poem is complicated. Let's read the poem closely listing the changes that we observe.

Observation 1: The first thing we notice is the change in punctuation. The paired dashes that set off and itemize the things the fist can hold, "Not power, not love, not gold," are dropped in stanza 4, but the change to the poem's meaning is imperceptible.

Observation 2: When we compare stanzas 2 and 5, another change occurs when the word "thought" is changed to "thing" in the fifth line of stanza 5. What is the effect of this change? Stanza 2 focuses on the relinquishing that takes place as death approaches, when *not a thought* is left for the mind. In stanza 5, the mind is again bereft of thinking and pain, but *not a thing* will remain. The emphasis is now on the physical, the tangible. The word "change" suggests a loss that goes beyond the relinquishing of thought; that is, the loss of the physical body, as well.

Observation 3: When we look at stanzas 3 and 6, we can see that the structural symmetry of the poem's rhyme scheme is still intact, but more words are changing. The "closing eye" becomes the "ardent eye"; the "resisting ear" becomes the "listening ear." In stanza 3, the dimmed sense organs are exhorted to respond, but in stanza 6, the same organs, when vital and strong, are seen as the *only* ones that can witness the thrush and his song.

Observation 4: Perhaps most dramatically, the mood of the verbs in the last three lines has changed. In this context, the word "mood" refers to the way a verb is used to indicate the speaker's attitude toward the subject. In stanza 3, the mood of the verb is imperative, as the unidentified speaker gives a command ("exhort the eye...to say"). In stanza 6, the mood of the verb is indicative, as if the speaker is stating a fact ("only the ardent eye...can say"). One interpretation of this change would be to say that in stanza 3 the speaker urges the diminishing senses (sight and hearing) to respond before it is too late, but in stanza 6, knowing that only ardent, listening senses can record the thrush's presence, and knowing that "not a thing" will be left, we are left with the impression that the speaker is referring wistfully, perhaps bitterly, to something that is no longer possible. One conclusion we could is that the poem is meant to encourage the speaker — and reader — to have an ardent eye and a listening ear, before "not a thing" is left.

As in James Weldon Johnson's poem, the structural symmetry in "Small Hands, Relinquish All" creates an expectation. The fact that stanzas 1 and 4 are virtually identical forces the reader to look closely for places where the two halves of the poem diverge. What is Millay's object in repeating and then breaking her own pattern, if not to invite a close, critical reading?

Earlier in this chapter, we talked about pattern poetry, where poets arrange words on a page so that the shape itself has significance. In some pattern poetry, from labyrinthine poems, to Herbert's "Easter Wings," the poem's shape is symbolic, and often, symmetrical. In twentieth-century pattern poetry we find shapes ranging from the representational "Concrete Cat" to the highly abstract "il pleut." Sometimes, however, we find a poem whose lines may have symmetrical properties but whose physical shape is almost invisibly interwoven with the ideas it expresses. Although such a poem makes use of symmetry, it is not usually considered to be a "pattern poem."

Working with left-justified margin, E.E. Cummings wrote a poem that appears nearly symmetrical when "folded" at its mid-point or tenth line.

lis
-ten

you know what i mean when
the first guy drops you know
everybody feels sick or
when they throw in a few gas
and the oh baby shrapnel
or my feet getting dim freezing or
up to your you know what in water or
with the bugs crawling right all up
all everywhere over you all me everyone
that's been there knows what
i mean a god damned lot of
people don't and never
never
will know,
they don't want

to
no

(*An Other E.E. Cummings* 93)

By interrupting the steady downward slope of the lines in the poem's second half, Cummings forces us to stop and pay attention to the repeated word *never*, in the 15th line where it stands alone. Cummings makes it impossible to ignore the fact that a lot of people will *never* know the suffering of those in war. The line is truncated, breaking the symmetry of the poem's shape. If we imagine the poem mirrored across the left margin, we would see a kite-like shape with the smooth descending shape of the lower half marred by the gap at line 15. Cummings understood well that interrupting the poem's symmetry, and repeating a single, adamant word, would jolt the eye and mind simultaneously.

Symmetry and Organic Form

In "Something," a poem whose shape moves with the rhythms of the sea's tides, the contemporary poet Mary Oliver uses symmetry to create a structure that seems organically suited to her subject. By using center-justification, Oliver allows the lines to expand and contract gently, as if being pulled. The enjambment or carrying over of ideas from one stanza to the next links the units of the poem together, calling to mind the "lace-mass" described in the poem's lines. Describing it mathematically, we could say that the lacy artifact of the poem is built by the motion created when the lines are reflected symmetrically across an imaginary axis drawn downward and passing through the mid-point of each line.

Something fashioned
this yellow-white lace-mass
that the sea has brought to the shore
and left —

like popcorn stuck to itself
or a string of lace rolled up tight,
or a handful of fingerling shells pasted together,
each with a tear where something

escaped into the sea. I brought it home
out of the uncombed morning and consulted
among my books. I did not know
what to call this sharpest desire

to discover a name,
but there it is, suddenly, clearly
illustrated on the page, offering my heart
another singular

moment of happiness: to know that it is
the egg case of an ocean shell,
the whelk,
which, in its proper season,

spews forth its progeny in such
glutenous and faintly
glimmering fashion, each one
chew and tearing itself free

while what is left rides to shore, one more
sweet-as-honey answer for the wanderer
whose tongue is agile, whose mind,
in the world's riotous plenty,

wants syntax, connections, lists,
and most of all names to set beside the multitudinous
stars, flowers, sea creatures, rocks, trees.
The egg case of a whelk

sits on my shelf in front of, as it happens, Blake.
Sometimes I dream
that everything in the world is here, in my room,
in a great closet, named and orderly,

and I am here too, in front of it,
hardly able to see for the flash and the brightness —
and sometimes I am that madcap person clapping my hands and singing;
and sometimes I am that quiet person down on my knees.

(*Why I Wake Early*, 38-39)

By arranging the poem's lines in a fluid, symmetrical shape, Oliver perfectly joins idea and form in this poem. As the speaker discovers the name and origins of the object she found on the beach, her words create another object — the poem itself — for the reader to discover and examine.

Symmetry as Metaphor

For a brief final look at symmetry in poetry, we turn to Fabio Doplicher whose poem "Asymmetry of the Universe" turns a theory from contemporary physics into a metaphor for the creative work of the poet. In an author's note accompanying the poem, Doplicher refers to the theory that the universe arose from "an *asymmetry* in the corresponding quantities of matter and anti-matter at an instant one thousandth of a second after the initial 'big bang'" (Doplicher 163). Doplicher describes his own writing within the context of that moment of ruptured symmetry. Using language that suggests symmetry and asymmetry ("imbalance," "mirror," "tightrope") he makes a connection between poetry and the creation of the universe.

> In a thousandth of a second
> the world was already made.
> In matter's imbalance, the void
> was born within the unknown lawgiver.
> Vessel of time spent, I've loved
> the necessity of verse, mirror of energy.
> You're not just word, my word.
>
> Today he who lies with you, poetry,
> Walks an odd tightrope of desire [. . .]
>
> (149)

The poem continues to develop the metaphor, and in stanza 9 the tension between symmetry and asymmetry becomes a powerful struggle for the artist:

> There is no correspondence, not even with the source,
> and we, the embittered gamblers,
> we bet all our chips on balance
> though true song needs divine disharmony.
>
> (153)

Despite our love of harmony and our desire for balance, Doplicher asserts that creative energy, the "true song" of the poet, requires "divine disharmony." In the asymmetrical moment when balance is overthrown, a universe — or a poem — may come into being.

Chapter 4 — Fractal Patterns

In Chapter 3 we extended classical geometric ideas to include hyperbolic geometry. In this chapter, we expand our conception of geometry even further with the introduction of *fractal geometry*, a branch of mathematics developed in the late twentieth century. We will also broaden our understanding of symmetry to include the idea of *symmetry of scale*, an integral part of this new geometry. *Fractals* are a contemporary example of a mathematical pattern whose abstract definition is made more accessible through intriguing visual representations. Many images in nature exhibit qualities similar to mathematical fractals, providing yet another venue for comprehending their complexity. The ideas of fractals have been connected to such diverse areas as the physical and natural sciences, economics and finance, psychology, the arts, image compression, literature, and, most importantly for our discussion, poetry.

Even if you have no idea how to define the term *fractal*, you have probably seen images that exhibit fractal qualities and heard references to fractals in popular media. Perhaps you heard the word used to describe the rough, jagged world of nature or to portray the complex repetition of forms in computer art. Or you may have encountered fractal-like images without knowing it. Possibly an image of a Buddhist mandala, a sacred design used to focus meditation, caught your imagination with its repeated interplay of circles and squares. Maybe you broke off a frond of a fern and remarked that this single piece was a miniature version of the entire fern, much like the resemblance between a single floweret and the entire head of a broccoli plant. Probably you've seen satellite images of a large river system and noticed that each branch replicates the entire river. But maybe you did not consider how similar this image is to one of lightning or the circulatory system in the human body. All three of these images have fractal qualities.

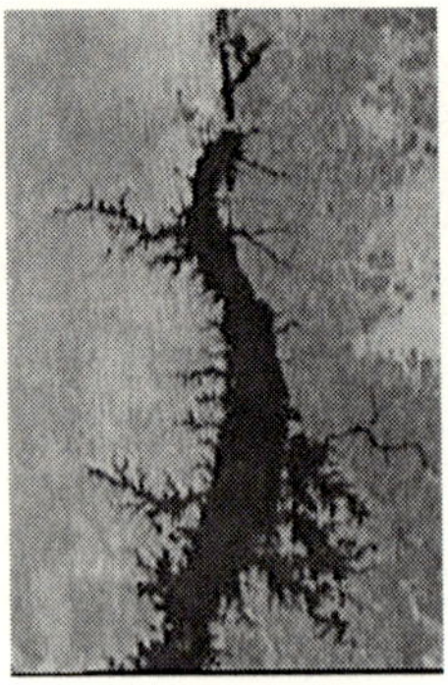

River system[8]
Figure 4.1

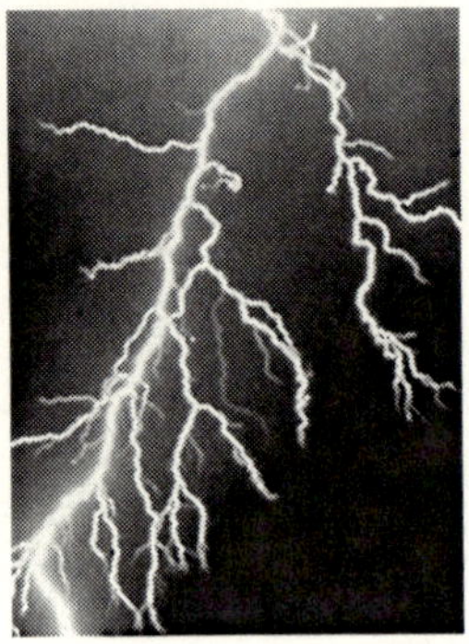

Lightning[9]
Figure 4.2

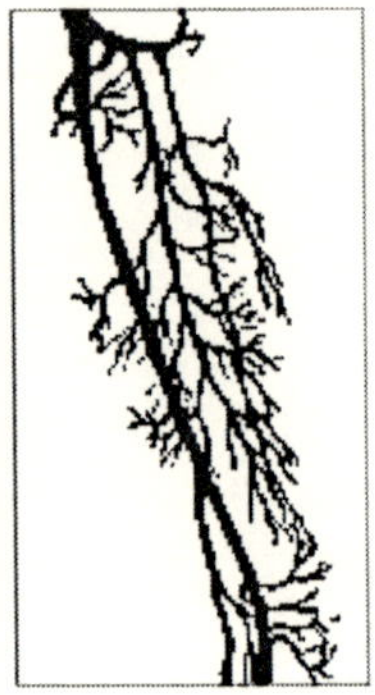

Circulatory System[10]
Figure 4.3

In fact, you may have even seen a picture of the most famous fractal image, the Mandelbrot Set (Figure 4.4), but perhaps did not know it had a name, a fascinating story, and properties that are quite remarkable. We will describe some of its history and attributes later in this chapter.

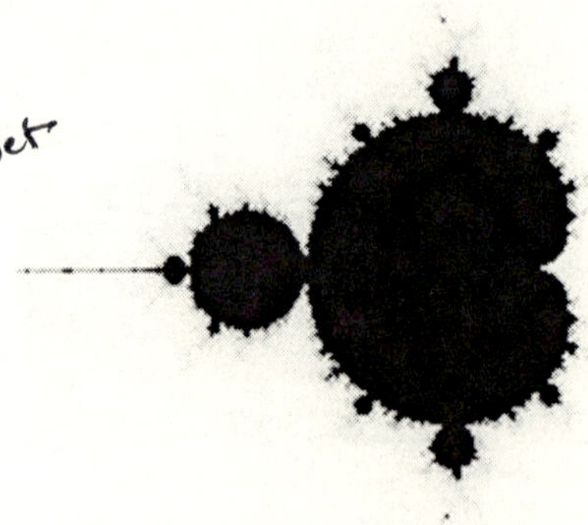

Figure 4.4

Any worthwhile discussion of fractals must include mathematical explanations, but this chapter is not intended to present all of the mathemati-

[8] Image from National Aeronautics and Space Administration, Copy-right free.
[9] Image from National Oceanic and Atmospheric Administration, Copy-right free.
[10] Image from National Institute of Diabetes and Digestive and Kidney Diseases, National Institute of Health, Copy-right free.

cal complexities of fractal geometry. Rather it will provide you with the important vocabulary of fractals, a basic mathematical explanation of how to create a simple fractal image, and many examples of images that exhibit fractal qualities. With this underpinning, we can then discuss fractals and poetry.

4.1 Basic Fractal Concepts

What exactly is a fractal? The definition of the term, or even a description of a fractal, varies according to the discipline in which it is being explained, with a visual artist using a different vocabulary from that of a mathematician. For example, a painter or photographer might say that a fractal image is an extremely complex work in which instances of either the entire work, or selected pieces of the work, are repeated. Perhaps each repetition is an exact replica of the original piece, or, more likely, the repetition just approximates the original. The picture below illustrates this idea through the shape, size, and placement of circles.

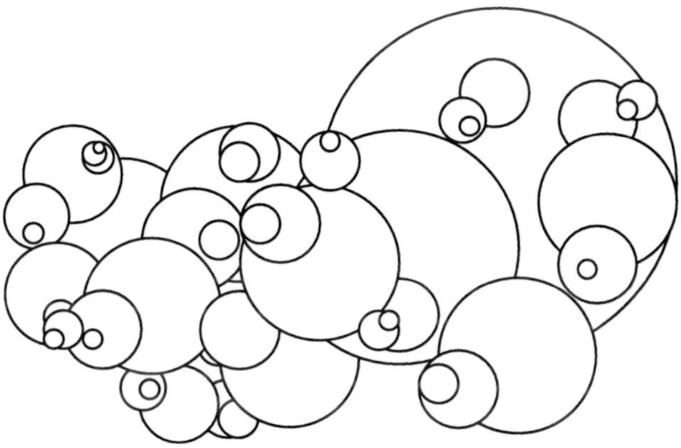

Figure 4.5

One could view Figure 4.5 as simply a collection of circles and say that the fractal nature of the drawing lies in the fact that the circles are all exactly the same, except for their size. However, on closer examination, it becomes clear that there are circle "families" appearing in clusters of three, as shown below in Figure 4.6.

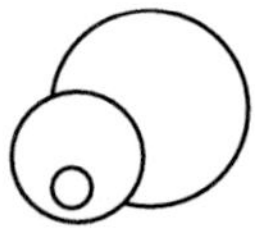

Figure 4.6

We can observe many such clusters of varying size, and the drawing suggests the presence of additional unseen clusters, some larger, some smaller. As we study the drawing, we can see that placement of the three-component circles is never exactly the same in any two "families." Rather than being identical, the family groups are approximate replications of each other.

When images are described as fractal, or more precisely "having fractal qualities," the artist or viewer is using terminology associated with an extensive body of mathematical discovery. Indeed, mathematicians have a precise vocabulary for defining and describing fractal images. Below is one definition that has mathematical rigor, without requiring a college course in measure theory. The vocabulary may at first be confusing, but each term is explained in the succeeding paragraphs. Notice the similarity to the artist's definition, but also observe how the vocabulary is more specific and the explanation more detailed.

> *A fractal is a mathematical object, such as a curve, that displays exact or approximate self-similarity on different scales. The image of a fractal is frequently obtained from a recursive or iterative formula and when magnified, the resulting detail looks surprisingly similar to the original object. Further, a fractal is generally irregular in shape and often has a dimension that is not an integer.*

Mathematical object - Mathematicians don't need to draw an object to know it exists or to use it in further mathematical study. Instead, the object may be represented by a computer algorithm, mathematical function, or formula. Some examples of such objects include a sequence of numbers, a set of points, a curve, or a surface.

Self-similarity - A mathematical object displays self-similarity if an image of the object contains smaller portions that replicate the entire object. These replicas may be "exactly similar," that is identical, except for the difference in size, or they may be "approximately similar," that is showing some differences or distortions beyond size, but preserving enough similarity to demonstrate the resemblance. For instance, in Figure 4.5 we see many examples of approximate self-similarity. It is easy to see differences both in the orientation of the circles, as shown in Figure 4.7, as well as in the relative sizes of the circles within a cluster, as seen in Figure 4.8.

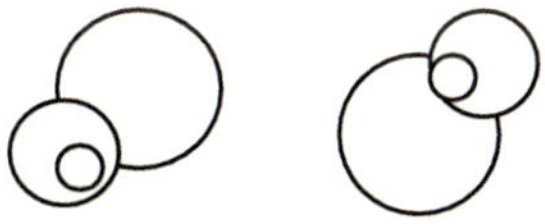

Difference in Orientation

Figure 4.7

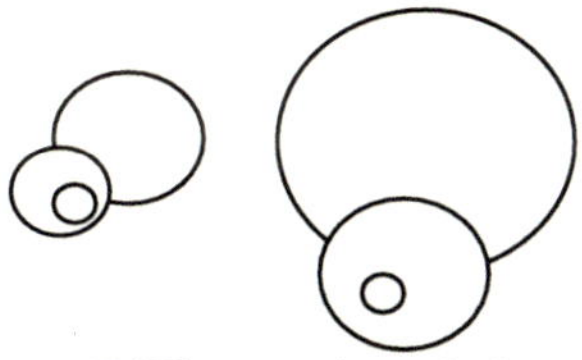

Difference in Relative Size

Figure 4.8

Different Scales - Magnification of a fractal image shows a new image similar to all or part of the original, but at a different scale or size. In Figure 4.9 we are looking at a portion of the right side of Figure 4.5, with shading used to illustrate scale. Vertical stripes highlight a large-scale version of our motif, while horizontal stripes delineate a midsized version. Two small-scale versions of the motif are shaded in polka dots.

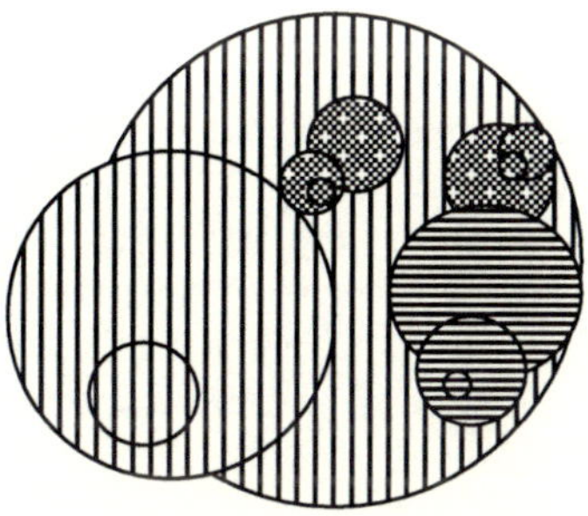

Figure 4.9

Recursive or Iterative process - This is the repeated application of a computer algorithm, mathematical function, or formula. One simple example of a recursive formula is $x_{n+1} = x_n + 3$. This formula leads to a sequence of values for x that depends upon the first value we choose for x. As shown in Figure 4.10, if the first x value is 0, the next four values of x would be 3, 6, 9, and 12. A different initial choice for x would lead to a different sequence of x values.

$$
\begin{aligned}
x_1 &= 0 \\
x_2 &= x_1 + 3 = 0 + 3 = 3 \\
x_3 &= x_2 + 3 = 3 + 3 = 6 \\
x_4 &= x_3 + 3 = 6 + 3 = 9 \\
x_5 &= x_4 + 3 = 9 + 3 = 12
\end{aligned}
$$

Figure 4.10

Ironically, many complicated mathematical fractals are produced from very simple iterative formulas like the one above.

Irregular shape - A fractal is usually not smooth, but rather displays jaggedness or roughness. For instance, a beach ball or an ice cream cone is not fractal in shape, but mountains or coastlines are said to have fractal shapes because of the jaggedness they display. If we keep zooming in on the edge of a beach ball, we still see a smooth surface. However, zooming in on a picture of the eastern coast of the United States reveals more and more detail. Bays and inlets, rock outcroppings and peninsulas each in turn display their own jagged inlets and outcrops.

Dimension - Dimension is one of the hardest notions to grasp about fractals. We know that a line is one-dimensional since it has length, but no width. A rectangle is two-dimensional, having both length and width, while a box is three-dimensional with length, width, and height. But what does it mean to say a fractal has dimension of 0.6 or 1.3? This unusual concept of non-integer dimension makes perfect sense to a mathematician. It relies upon a different way of defining dimension, one that is related to self-similarity and scale. We will discuss this topic later in this chapter.

Viewing the photograph of an actual shoreline helps to illustrate the two concepts of irregular shape and dimension. The photograph in Figure 4.11 shows Yellowstone Lake as seen from the Storm Point Trail in Yellowstone National Park, Wyoming.

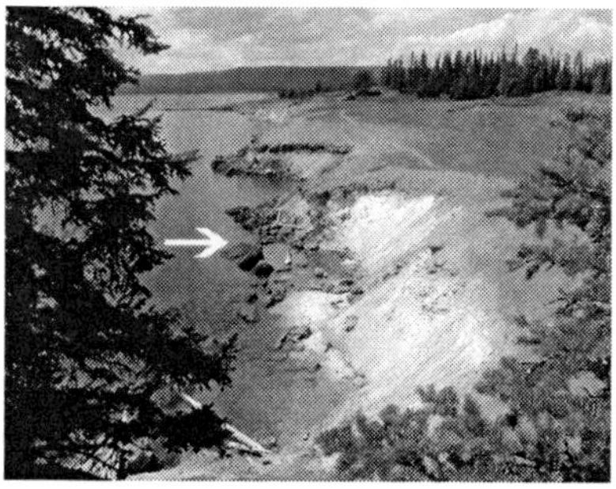

Figure 4.11

The land at the lake's edge is rugged, showing a number of large inlets, but it is difficult to grasp just how rough this section of the shoreline is. As we zoom in at the white arrow on Figure 4.11, we obtain the more detailed view seen in Figure 4.12. We can see smaller inlets and outcrops within the larger semicircular inlets.

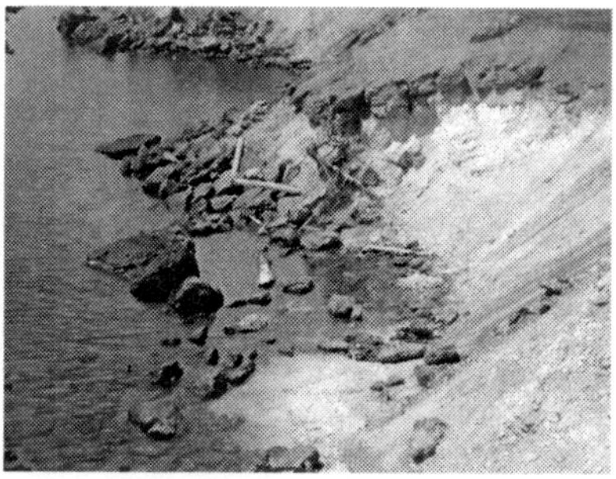

Figure 4.12

Zooming in ever closer to the shore illustrates even more details of jagged rocks forming smaller and smaller inlets. The arrow in Figure 4.13 points out one of these tiny inlets.

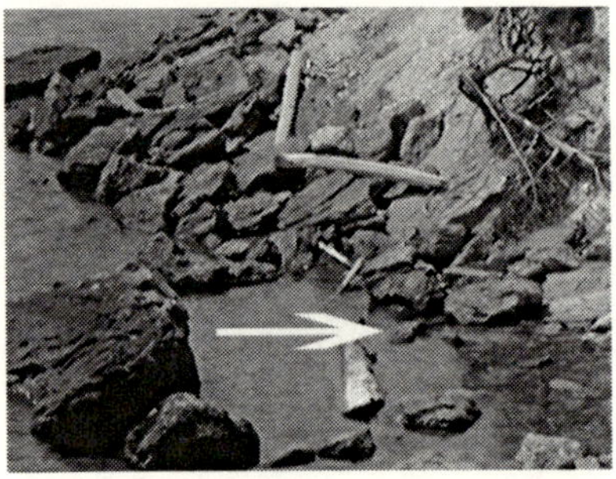

Figure 4.13

What is the dimension of such a rough shoreline? Is it one-dimensional, like a line? Two dimensional, like a rectangle? When the jaggedness is so irregular that the shore "line" starts to resemble something that is infinitely jagged, a mathematician would say that the dimension of the shore lies somewhere between our familiar dimensions of 1 and 2.

Fractals in Nature

Fractal patterns appear throughout our world across multiple scales, from the minute to the global. The tiny pattern of frost on a window (Figure 4.14), has been enlarged five times to show the detail of self-similarity, while no magnification is needed to see the fractal-like pattern of damage left by an ice storm (Figure 4.15).

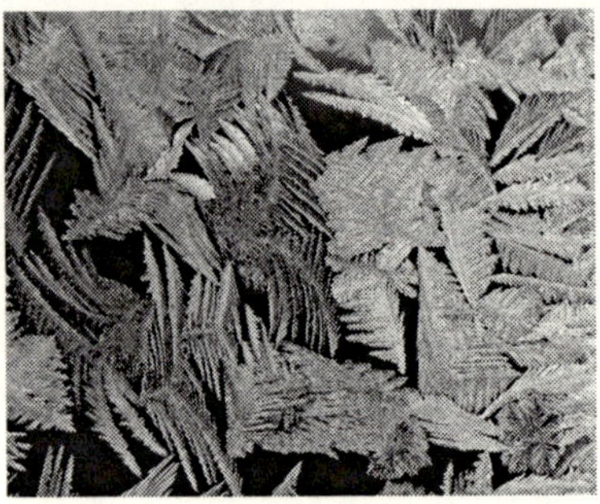

Figure 4.14

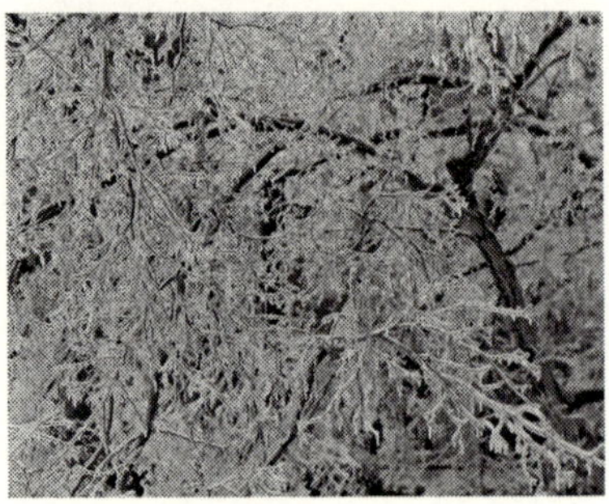

Figure 4.15

Some other examples of nature photos exhibiting self-similarity, differing scales, jaggedness, and irregular shape are shown below. The first is a cyano-bacteria mat at the base of a geyser at Mammoth Hot Springs, Yellowstone National Park. Miniature terraces are formed by the runoff of bacteria.

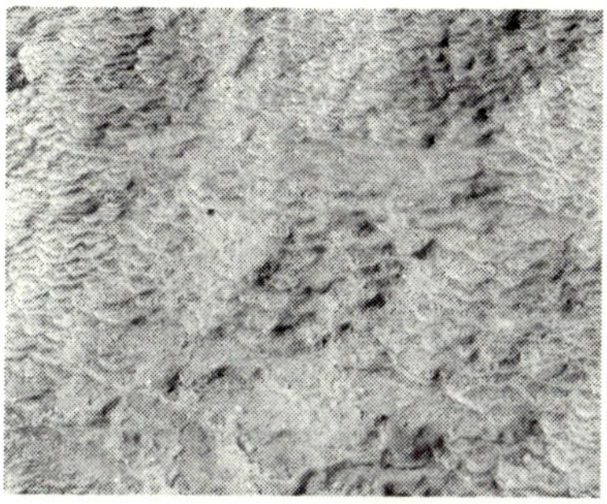

Figure 4.16

The next two photographs echo the "steps" seen above in the bacteria. They show how the mineral travertine forms a similar, but larger scale, terraced landscape at Minerva Terrace, an extinct geyser at Mammoth Hot Springs in Yellowstone National Park.

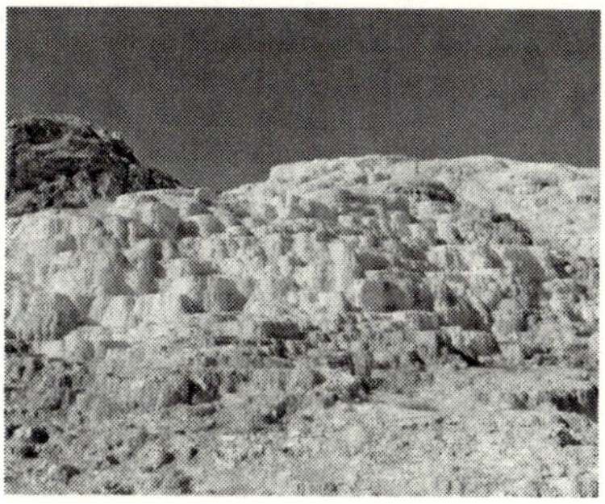

Figure 4.17

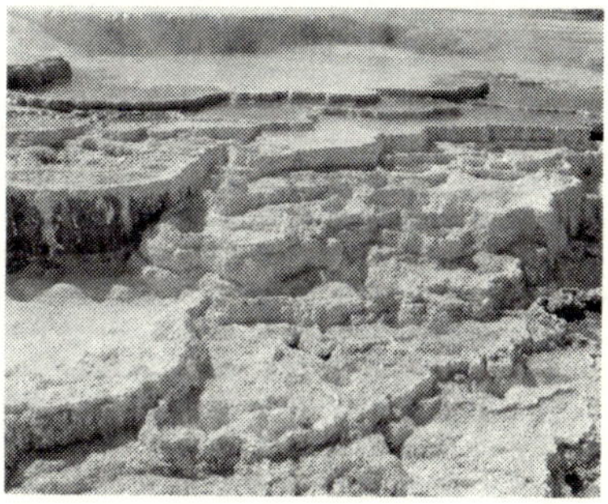

Figure 4.18

Lastly, in Figure 4.19 and 4.20 we see the peaks of the Grand Teton Mountains taken from different vantage points, one distant, the other near the base of the range.

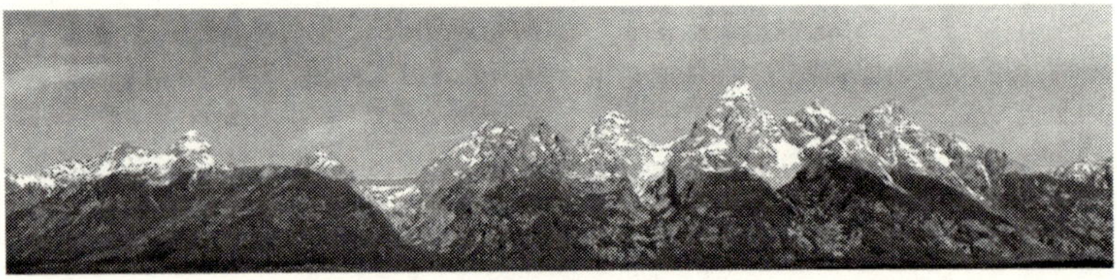

Figure 4.19

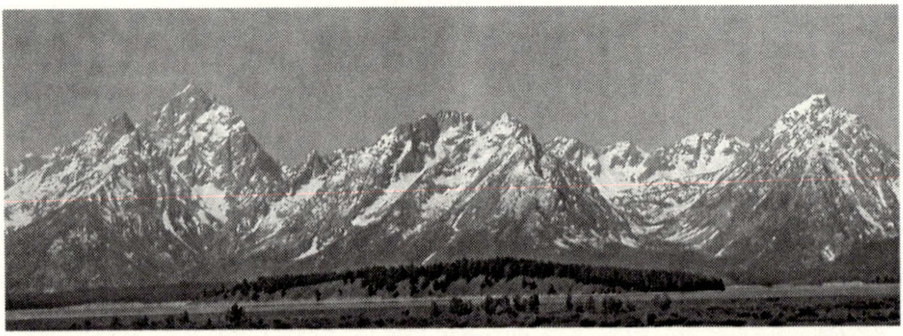

Figure 4.20

The detail in the closer shot (Figure 4.20) magnifies each peak of the distant view (Figure 4.19). With enlargement, we are able to discern the smaller scale versions of the jagged peaks in the Tetons, just as the smaller inlets of the shoreline emerged under magnification in Figures 4.11 through 4.13.

For the mathematician, these patterns of scale, repetition, and roughness are both aesthetically pleasing and geometrically satisfying. But where does the idea of fractals come from? Let's answer the question by turning our attention to the history of fractals and the mathematics behind creating fractal images.

4.2 A Brief History of Fractal Geometry

Natural forms that exhibit fractal qualities have existed as long as there have been mountains, plants, and rivers, but the terms *fractal* and *fractal geometry* are modern words dating to the 1970s. Benoit Mandelbrot, a Polish-born mathematician who was educated in France and emigrated to the United States in 1958, coined the word "fractal" to describe the new geometry he had developed. Mandelbrot based the word fractal on the Latin word "fractus" meaning fragmented or broken, and used it in print for the first time in his 1975 book, *Les Objets Fractals: Forme, Hasard, et Dimension.* Mandelbrot then developed the links between abstract mathematical concepts and fractal images we see in the natural world in his 1982 work *The Fractal Geometry of Nature.* Mandelbrot's work was related to theoretical work done by mathematicians in the late nineteenth and early twentieth centuries, but he used the processing power of modern computers to iterate a simple mathematical formula and produce the beautiful image now known as the Mandelbrot Set. He tied together this theoretical, analytical, and computational work in a new geometry he called "fractal geometry."

Let's examine some examples from the work of mathematicians in the late nineteenth and early twentieth centuries who inspired Mandelbrot. Mathematicians such as Giuseppe Peano, David Hilbert, Helge von Koch, Waclaw Sierpinski, and Georg Cantor produced shapes and curves that today we identify as fractals. But these mathematicians used descriptions such as *space-filling curves*, *iterative removal*, and *replacement* when describing their work. The shapes they created exhibited such puzzling mathematical concepts that they defied conventional geometric explanation. It wasn't until Benoit Mandelbrot gave us a vocabulary to discuss these important curves that they became widely known beyond elite mathematical circles.

Mathematical Thought Before Mandelbrot

Giuseppe Peano was the first to describe a "space-filling curve" using a mathematical formula. Through an ongoing series of iterations, he showed that a simple one-dimensional curve could be altered to completely fill a two-dimensional space. In 1890, Peano described analytically how such a curve would be formed but did not actually draw the curve. We can create a representation of Peano's Curve as shown in Figures 4.21A - 4.21C. Figure 4.21A is made up of 9 short line segments. If we replace each of these line segments with a small-scale version of Figure 4.21A, we obtain Figure 4.21B, composed of 81 short line segments. Performing another iteration of replacement on the 81 line segments produces Figure 4.21C. If this were kept up indefinitely, the curve would fill a solid square.

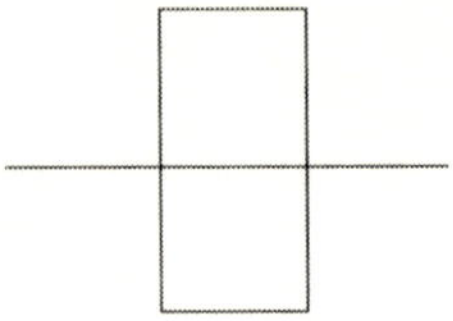

Figure 4.21A

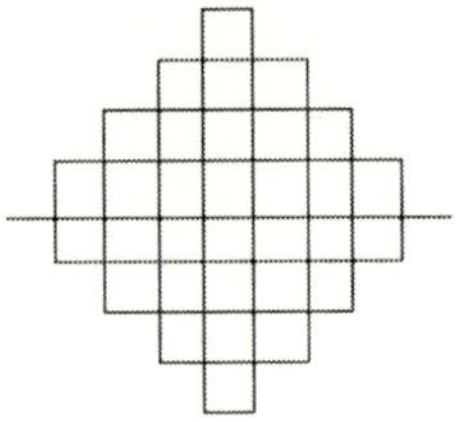

Figure 4.21B

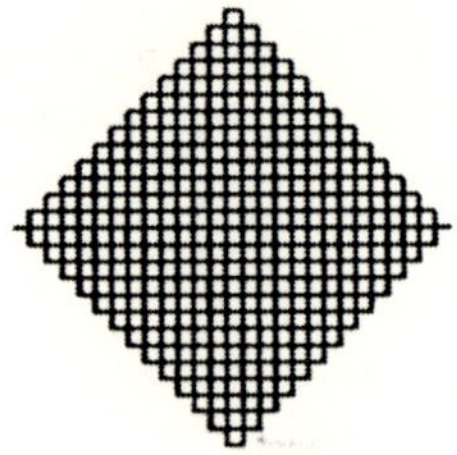

Figure 4.21C

It was the dimension of Peano's curve that so astounded mathematicians at the end of nineteenth century. We would expect a curve to be one-dimensional (having length, but no width), but Peano's Curve defied logic by filling a two-dimensional space. Today we understand the paradox in terms of a new type of dimension, namely fractal dimension, more formally called Hausdorff-Besicovitch Dimension. The Hausdorff or fractal dimension of a self-similar fractal is given by:

$$Dimension = \frac{Log(number\ of\ self\text{ - }similar\ pieces)}{Log(magnification\ factor)}$$ [11]

In the Peano Curve there are 9 self-similar pieces or short line segments in the original drawing. Each segment would need to be magnified by 3 to obtain one complete horizontal line segment. Thus the dimension of the Peano Curve is 2, as shown below:

$$\frac{\log(9)}{\log(3)} = \frac{\log(3^2)}{\log(3)} = \frac{2\log(3)}{\log(3)} = 2$$

Just one year after Peano described his curve, David Hilbert constructed an example of such a two-dimensional curve. He used the shape shown in

[11] A more complete explanation of this formula is given by Robert L. Devaney in *Chaos, Fractals, and Dynamics: Computer Experiments in Mathematics*. Menlo Park, CA: Addison-Wesley Publishing Co., 1990, 145-6.

Figure 4.22A as his basic unit of construction. He reduced the shape in size and produced four identical shapes linked by straight lines to create the image in Figure 4.22B. In each successive panel, the previous shapes are reduced, some are turned, and then linked by straight edges to produce the maze-like effect seen in Figure 4.22D.

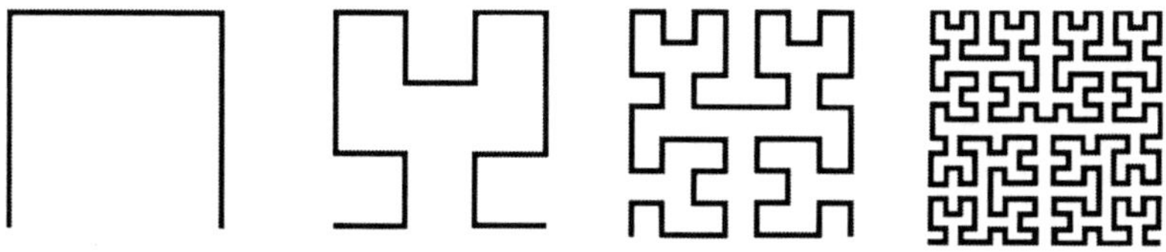

Figure 4.22A Figure 4.22B Figure 4.22C Figure 4.22D

These two examples show curves that are one-dimensional, but fill a two-dimensional space when the iteration is continued infinitely. This paradox of a one-dimensional object mutating into something two-dimensional was so unsettling to mathematicians that they labeled such curves as "pathological" or "monster curves." Such curves would not be accepted as "natural" until new ideas in measure theory were developed by Felix Hausdorff and later built upon by Abram Besicovitch. The work of Hausdorff and Besicovitch in the first half of the twentieth century provided a foundation for the work of Benoit Mandelbrot.

Toward the end of the nineteenth century and into the early part of the twentieth century, other shapes with strange dimensions were studied. The most famous include Cantor Dust, produced by Georg Cantor in the late 1800s; the Koch Snowflake, first published in 1906 by Helge von Koch; and the Sierpinski Triangle, developed by Waclaw Sierpinski in the 1920s. These shapes, which we now call fractals, are produced through successive steps of replacement or removal. When the iteration of replacement or removal is done infinitely, the results can be quite startling. Let's turn our attention Koch, Sierpinski, and Cantor's work.

The Koch Snowflake is produced by iteratively removing one third of each side of a triangle. First we draw an equilateral triangle, as shown in Figure 4.23.

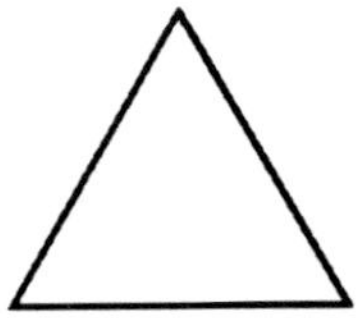

Figure 4.23

In the first removal/replacement step, the middle third of each side of the triangle is removed and replaced with an equilateral triangle whose side is 1/3 the length of the original.

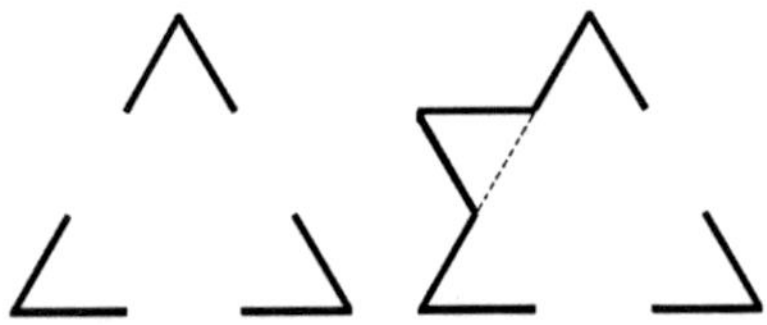

Figure 4.24

When this is done to all three sides of the original triangle, we get a star-like figure as shown in Figure 4.25.

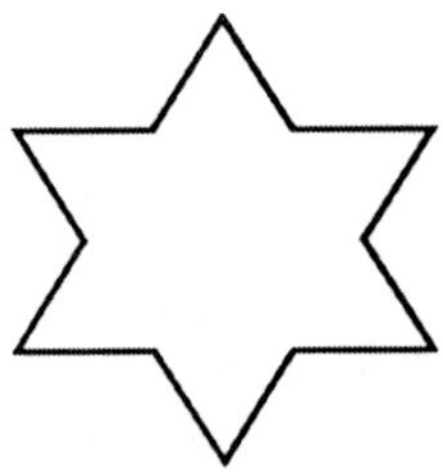

Figure 4.25

Again and again, the middle third of each side of the figure is removed and replaced with a triangle 1/3 the length of the figure's sides. The process produces a series of shapes that start to resemble a snowflake, as seen in Figures 4.26A - 4.26C.

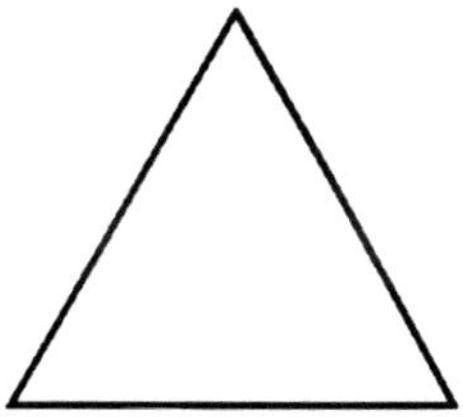

Figure 4.26A

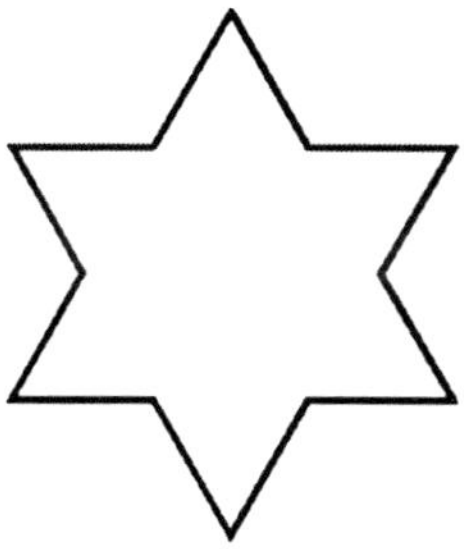

Figure 4.26B

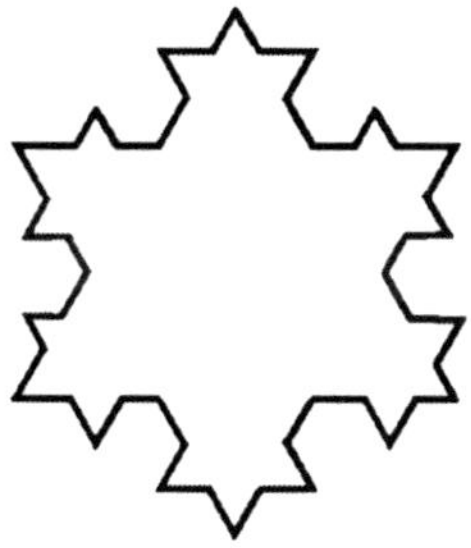

Figure 4.26C

One remarkable thing about the Koch Snowflake is that the figure has a finite area, but its perimeter (or outside edge length) approaches infinity. Its dimension is equally as unpredictable. In fact, it is not an integer value. We can determine the dimension of the Koch Snowflake by following the earlier example of the Peano Curve. In the first iteration of the Koch Snowflake we see that 4 smaller pieces replace each side of the original triangle. Also, each of these smaller pieces would have to be magnified 3 times to produce the original side. The dimension D of the Koch Snowflake is based on these two numbers, 4 and 3, and is:

$$D = \frac{\log(4)}{\log(3)} \approx 1.262.$$

Thus, the Koch Snowflake is neither one-dimensional nor two-dimensional, but rather lies between these conventional dimensions.

Like Helge von Koch, Waclaw Sierpinski produced the famous Sierpinski Triangle by iterative removal. To make a Sierpinski Triangle, we begin with a black, solid-colored, equilateral triangle, as shown in Figure 4.27A. The midpoints of each side of the original triangle are joined to form an inner triangle that is then removed (Figure 4.27B). Three black equilateral triangles remain, each having a side 1/2 the length of the original triangle. As we continue to remove the middle third of each of the black triangles, a "lacey" figure starts to emerge that contains less and less black area.

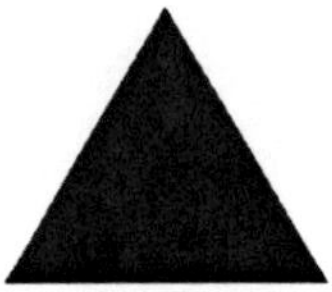

Figure 4.27A

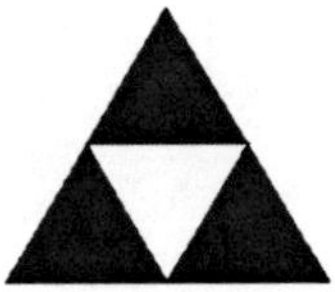

Figure 4.27B

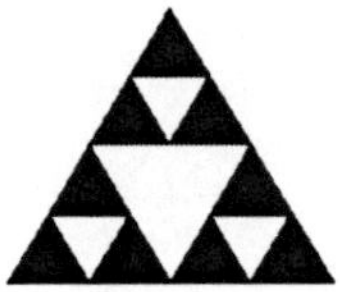

Figure 4.27C

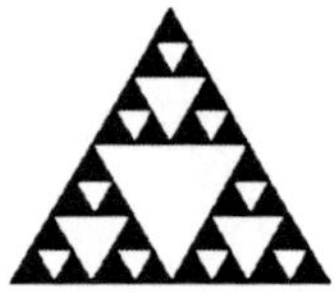

Figure 24.7D

Since the original triangle is replaced by 3 exact replicas, each 1/2 the size of the original, the dimension of the Sierpinski Triangle, is:

$$D = \frac{\log(3)}{\log(2)} \approx 1.585.$$

The perplexing thing about this fractal is that if we indefinitely remove the middle of each solid-colored triangle, the remaining figure would have a perimeter that approaches infinity, while its area would approach zero. The ever diminishing area after each iteration accounts for another name for this fractal — the Sierpinski Sieve.

For a third example of iterative removal, let's examine how Georg Cantor created the fractal image known as Cantor Dust, found below in Figure 4.28. We begin by drawing a line of some fixed length. An identical second line is drawn below the first, but the middle third of the second line is removed. This iteration is continued, and, with each successive row, the middle third of each line in the previous row is removed. Eventually, the line dissolves into mere specks, or Cantor Dust. Very rapidly the "dust" becomes so small we cannot see it with the naked eye, and yet it continues to be arranged in the same pattern as at the beginning of the removal process. We have only removed a third of a line each time, but we are left with seemingly nothing!

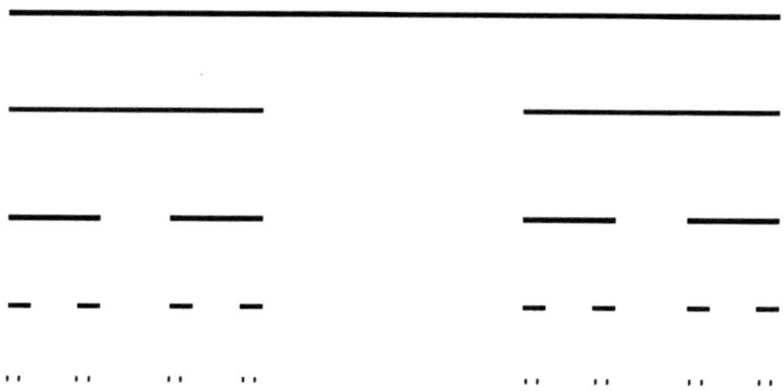

Figure 4.28

The shapes that Peano, Hilbert, Koch, Sierpinski, and Cantor produced, without the aid of computers, had all been created by the 1920s, although they were not referred to as fractals until the 1970s. At every magnification, a portion of each figure exhibits the same properties as the entire figure, showing us exact self-similarity over a variety of scales. In each case, a very simple process is used to create a highly complex figure.

Julia Sets and the Mandelbrot Set

Another important name in the history of fractals is Gaston Julia, a mathematician born in 1893, who produced an award-winning paper at the age of twenty-five on the iteration of rational functions in the complex plane. His discussion was theoretical and included no images. Today, we say that Julia's iterative functions produce Julia Sets, and we use powerful computers to draw the representations of these sets. But Julia never drew a Julia Set; in fact, he probably never imagined a Julia Set as anything but a theoretical object. With the help of Mandelbrot's computer imaging in the 1970s, the theoretical work of Julia was made visual. Julia Sets and the Mandelbrot Set are not produced through simple iterative replacement or removal, as in the fractals of Peano, Hilbert, Koch, Sierpinski, and Cantor, but rather through examining the behavior of Complex Numbers under recursive functions. Julia Sets and the Mandelbrot Set do not exhibit exact self-similarity, like Figures 4.21 - 4.28, but rather are examples of fractal images with approximate self-similarity.

To fully understand how Julia and Mandelbrot Sets are created, we need to know how to plot points in the Complex Plane. Recall that a Complex Number is of the form $a + bi$ where a and b are Real Numbers and $i = \sqrt{-1}$. Thus each Complex Number is the sum of a Real component (a) and an Imaginary component (bi). We use the x-axis for plotting the Real portion of each number and the y-axis for plotting the Imaginary portion. Below you will see some numbers plotted in the Complex Plane. For example, the number $3 + 2i$ is plotted by moving three units to the right of the origin on the Real Axis and then 2 units up in the direction of the Imaginary Axis.

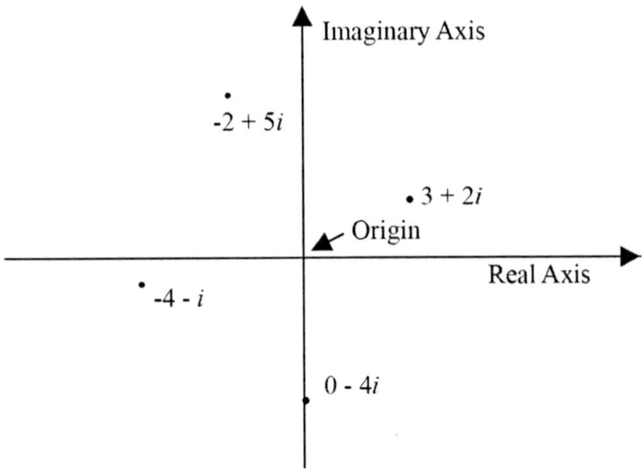

Let's examine some images of Julia Sets, pictured here as the points in the Complex Plane that are not colored black. There are as many Julia Sets as there are numbers in the complex plane. Each Julia Set is associated with a "seed" or a particular complex number that creates it. Three different Julia Sets and their respective seeds are shown in Figures 4.29 - 4.31[12].

[12] All Julia Set images produced using the software *Julia's Dream*.

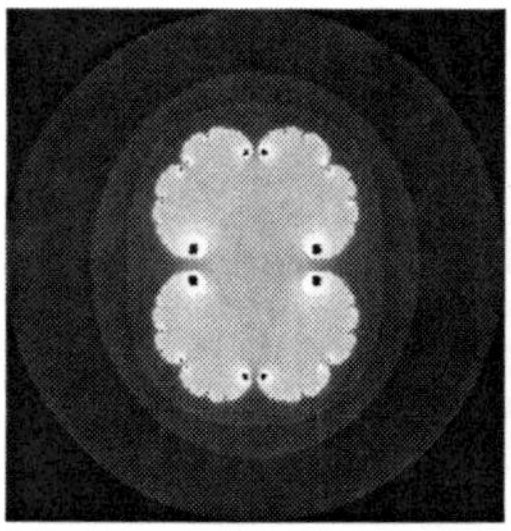

Seed = $0.27 + 0.0i$

Figure 4.29

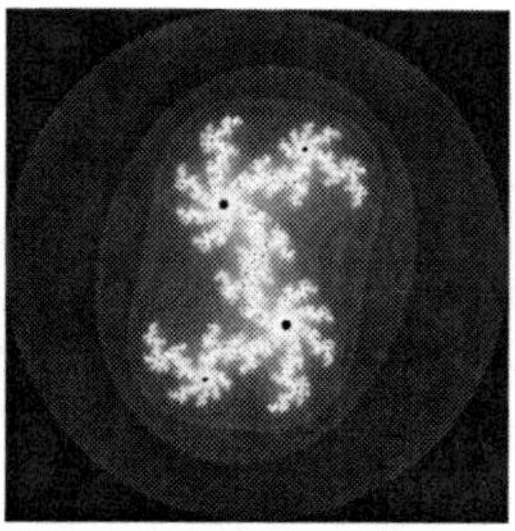

Seed = $0.25 - 0.25i$

Figure 4.30

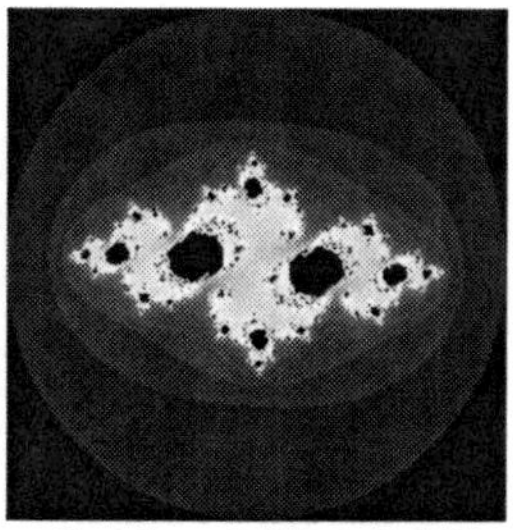

Seed = $-0.76 + 0.16i$

Figure 4.31

> To create a Julia Set, select a complex number C as a seed for the iterative function $Z_n = Z_{n-1}^2 + C$. For the selected value of C, allow Z to take on all values of complex numbers $x + iy$ in the complex plane. A complex number Z will belong to the Julia Set of C if, when it is iterated in the function, the resulting sequence of values remains bounded. If the iteration causes the sequence of Z_n to grow infinitely large under the iteration, then Z is not in the Julia Set of its seed C.

Mandelbrot revived interest in Julia Sets by using the computer to visualize the patterns that these iterated functions produced. Mandelbrot's own creation, the Mandelbrot Set, is closely linked to Julia Sets and is formed by iterating the same function $Z_n = Z_{n-1}^2 + C$, but with a slight twist. Instead of fixing the value of C and producing a different Julia Set for each seed, the Mandelbrot Set establishes the initial value of Z as zero and iterates the function for every number C in the complex plane. As in Julia Sets, bounded sequences are used to determine if a given value of C is in the Mandelbrot Set. If a value of C is in the Mandelbrot Set, it is colored black, producing the famous image seen in Figure 4.32.[13]

[13] All Mandelbrot Set images produced using *Aros Fractals*.

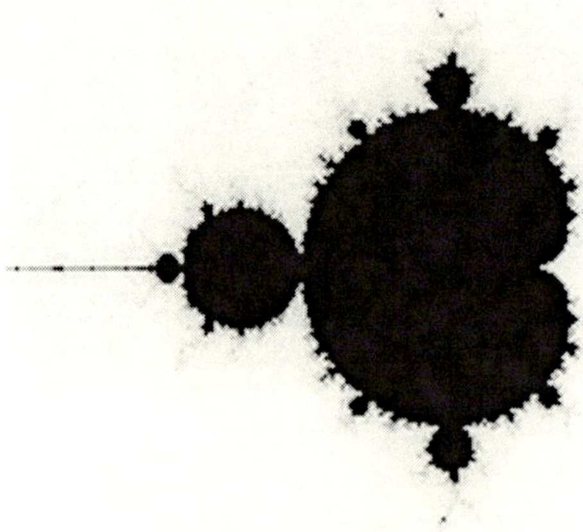

Figure 4.32

To create the Mandelbrot Set, use the iterative function:

$$Z_n = Z_{n-1}^2 + C$$

where Z is a complex number of the form $x + iy$.
Begin with $Z_0 = 0$ (the seed of the Mandelbrot Set). Then we see that:

$$Z_1 = C$$

$$Z_2 = C^2 + C$$

$$Z_3 = (C^2 + C)^2 + C$$

$$Z_4 = ((C^2 + C)^2 + C)^2 + C$$

If we continue this iteration, we create an infinite list or sequence of values: Z_0, Z_1, Z_2, Z_3 . . . This sequence is called the orbit of Z_0 for any value of C. If a given value of C creates an orbit that is bounded, then the point C is in the Mandelbrot Set and we color the point black. If a given value of C creates an orbit that becomes infinitely large, then C is not in the Mandelbrot Set. Values of C whose orbits keep increasing their distance from the origin are colored according to how quickly the orbit moves away from the origin. The grey, antenna-like fringes of the set represent points in the complex plane that are on the edge of the Mandelbrot Set. They are often referred to as "escaping" slowly from the origin under iteration. This recursive process of determining the behavior of the orbit of Z_0 for all values C in the complex plane creates the fractal known as the Mandelbrot Set.

We will examine the Mandelbrot Set carefully, zooming in on various parts, to show the approximate self-similarity that is exhibited. First we need a vocabulary to discuss various shapes found within the Mandelbrot Set. On a large scale we can see a main shape called the "cardioid," so named because of its heart shape, as well as shapes referred to as "bulbs" and "antennae."

Cardioid

Figure 4.33

Bulb

Figure 4.34

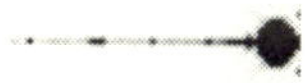

Antenna

Figure 4.35

The strange qualities of the Mandelbrot Set are best explored by zooming in on different parts of the set to examine instances of self-similarity. There are many bulbs surrounding the main cardioid, including a large bulb on the left of the cardioid and two medium-sized bulbs at the top and bottom. We will focus closely on the bulb at the very top of the Mandelbrot Set, designated by the arrow in Figure 4.36.

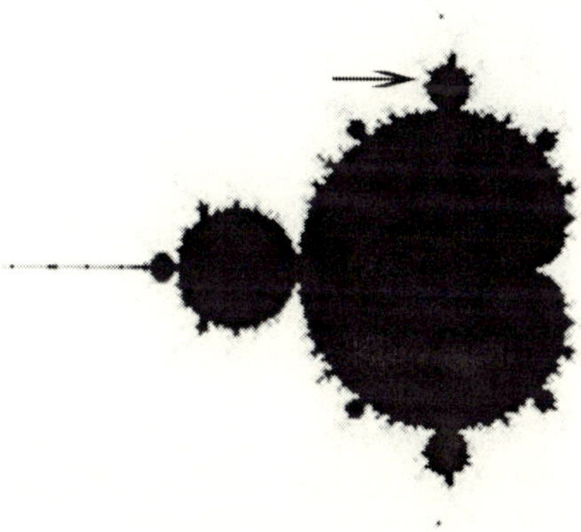

Entire Mandelbrot Set

Figure 4.36

What is not obvious on first glance at Figure 4.36 is that the bulb we are studying is surrounded by images of similar bulbs. Zooming in at the arrow of Figure 4.36 shows us these smaller bulbs in Figure 4.37.

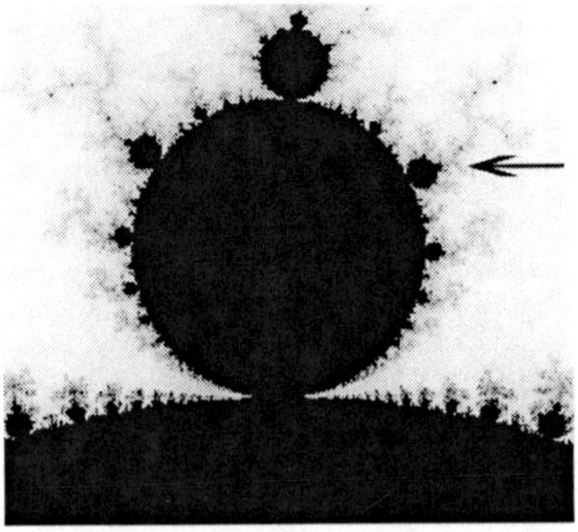

Figure 4.37 (First zoom)

We will magnify this image a second time, focusing on the arrow in Figure 4.37 to produce the enlargement seen in Figure 4.38.

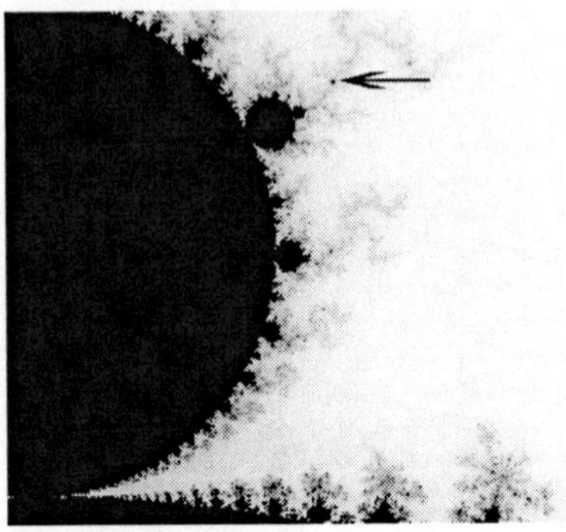

Figure 4.38 (Second zoom)

There is a tiny speck just off the edge of the bulb in Figure 4.38 where we can magnify a third time and see something quite amazing. On enlargement, that seemingly insignificant speck strongly resembles the entire Mandelbrot Set, as shown in Figure 4.39. It is not exactly self-similar, but it is approximate enough that we can recognize the main cardioid, surrounding bulbs, and long antenna.

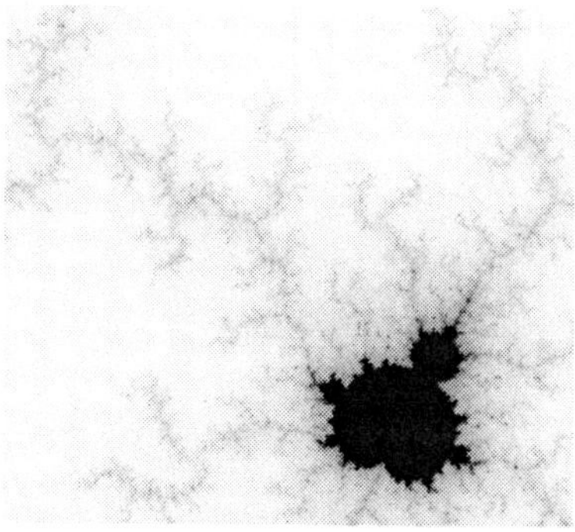

Figure 4.39 (third zoom)

The top bulb of the original Mandelbrot Set is not the only place in which we can find approximately similar images of the entire set. If we focus on the portion of the set known as the main antenna, we will find many approximate repetitions of the entire Mandelbrot Set. We begin to zoom in on the antenna in Figure 4.40 at the arrow.

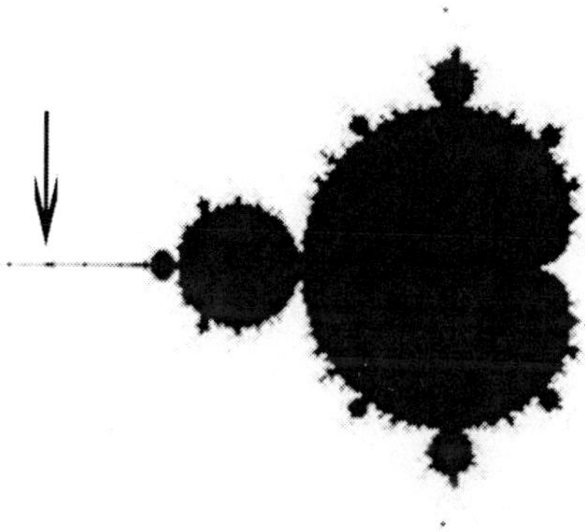

Entire Mandelbrot Set

Figure 4.40

In the next image, Figure 4.41, we see what appears to be another replication of the entire set with a small dot to its right (indicated by the arrow).

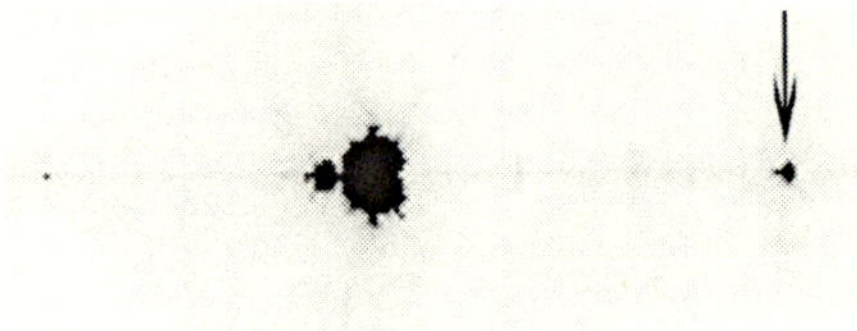

Figure 4.41 (first zoom)

Zooming in once again at the arrow of Figure 4.41 we obtain Figure 4.42, yet another approximate replication of the entire Mandelbrot Set.

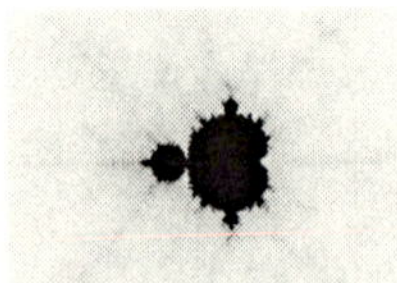

Figure 4.42 (second zoom)

Clearly there is a close connection between Julia Sets and the Mandelbrot Set since both sets are formed by the infinite iteration of the same simple equation in the complex plane. In fact, if a point *inside* the Mandelbrot Set is used as the "seed" or starting value for the recursion in a Julia Set, the image that is created appears somewhat solid or what mathematicians call "connected."[14] In contrast, if a point *on the edge* of the Mandelbrot Set is chosen for the seed, then the corresponding Julia Set is less solid and appears more feathery. Finally, if the seed for the Julia Set is chosen from *strictly outside* the Mandelbrot Set, the Julia Set appears dust-like or "disconnected."

Below are two images of the Mandelbrot Set (outlined in white) superimposed over an accompanying Julia Set (in gray). Each Julia Set was created from a seed located at the arrow. In the left drawing, Figure 4.43, the seed is actually just inside the Mandelbrot Set at the cusp of the cardioid. This leads to a connected, almost solid gray area representing the Julia Set. In the drawing on the right, Figure 4.44, the seed is located on the outside edge of one of the Mandelbrot bulbs, leading to a fairly disconnected Julia Set that almost resembles snowflakes.

---

[14] The term "connected" has a precise meaning in mathematics, but for the general reader the term may be thought of as meaning "all of one piece" or not broken apart.

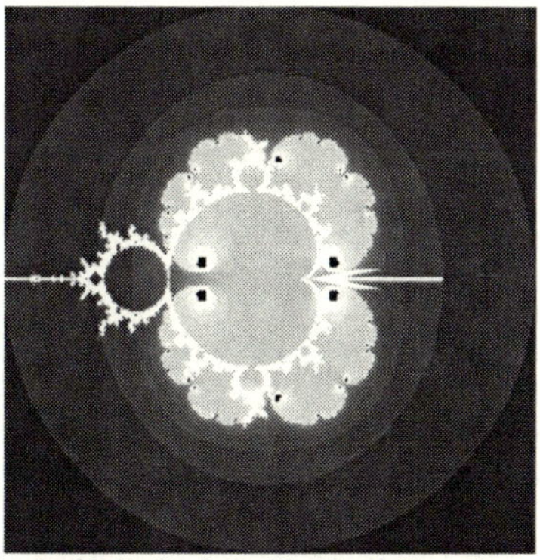

Seed of Julia Set is Inside Mandelbrot Set

Figure 4.43

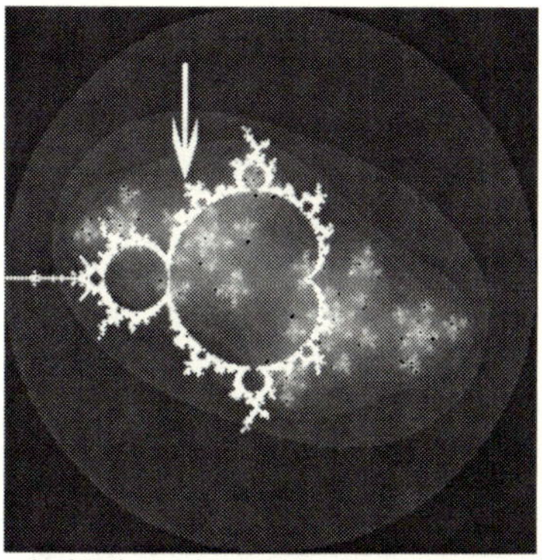

Seed of Julia Set is at Edge of Mandelbrot Set

Figure 4.44

Chaos and Fractals

Our history of fractals would not be complete without a brief discussion of the fractal's close relatives — chaos and dynamical systems. The field of dynamical systems, a relatively new area of study for mathematicians, relies on the power of computers to compute and visualize the patterns of nonlinear processes that change over time. Shifting weather patterns, fluctuating stock prices, the motion of planets, and turbulent liquid flows are all examples of dynamical systems that exhibit chaotic or unpredictable behavior. Such a system may be simple to describe mathematically, but the iterative nature of the system can produce extremely complicated results. Often, a small change to the initial conditions of the system will have a profound effect. Where do fractals come in? Fractals may be found in the images or visualizations of dynamical systems. Fractals themselves also exhibit chaotic behavior. The connections between fractals and chaos are complex and beyond the scope of this book, but it is important to point out that many mathematicians, scientists, engineers, and computer scientists have taken a team approach to the study of dynamical systems, chaos, and fractals. Some of these interdisciplinary groups also include visual artists and musicians, along with literary scholars, writers, and poets, all drawn to the ideas of this new science.

4.3 Fractals and Poetry

> [. . .] What do you want?
> Patterns that make you utter
> *surprise* — nonlinear
> plots, fractal
> repetitions, Mandelbrot sets,
> a template
> in the chaotic
> penetralia [. . .]
>
> (Holmes 280)

Artists, photographers, and writers have all been excited by fractals, and with good reason. It is fascinating to think that elaborate repeating patterns exist virtually unseen in the world around us. When creative individuals outside the scientific laboratory first learned about fractals, usually by seeing colorful computer-generated images, they began to look for ways to apply that knowledge to their own work.

In many ways, fractals have given readers and writers a new set of lenses through which to consider poetry. Contemporary poets have incorporated fractal patterns into the arrangement of words on a page and the use of language within a poem. Some have explored qualities of fractals in the ideas or themes of a poem. Poets and literary scholars have also coined the term "fractal poetry" and have taken fractal principles into account when reading and analyzing poems written long before Benoit Mandelbrot was born.

Fractal Images in Words

Sometimes a poet's knowledge of fractals is obvious from the way a poem looks on the page. As we saw in Chapter 3, humans have for centuries created *pattern poetry*, where the physical arrangement of words on a page incorporates deliberate visual elements. Fractals have added yet another possibility for bringing together words and visual images.

In his poem "The Cantor Dust" Rodrigo Siqueira uses clever visual word play to display the properties of the fractal:

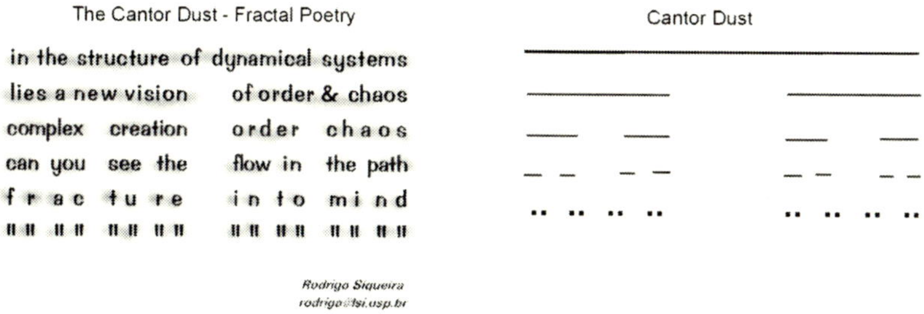

Figure 4.45

While the poem's first line invites us to think about what lies within dynamical systems, its form suggests the pattern of the Cantor Dust fractal (as seen in Figure 4.45). If this poem were to continue in another line (we might think of it as the next level of iteration), the words would, theoretically, become even smaller and more fragmentary, as a portion of the characters would again be removed. Although the poet does not remove exactly one-third of each line, as is done in when creating the Cantor Dust fractal, the consistent removal of approximately one-third of the characters in each succeeding line mimics the process used to create the fractal. What would happen if the poem were to continue? Would the words disintegrate? The poem's representation of the fractal teases us into considering the possibility of more lines, invisible ones, existing only in the reader's imagination. Siqueira has united the visual and theoretical properties of the Cantor Dust pattern in a poem whose form and content both refer to fractals.

In addition to poems where the physical arrangement of words on the page suggests a fractal shape, there are many instances where poets reveal a knowledge of fractals in the language they use. This use of fractals may be metaphorical or may extend into the structure of the poem itself, when a poet employs patterns of phrasing that may be described as fractal.

The Mandelbrot Set and the Language of Poetry

In the language and structure of several contemporary poems, we can see obvious references to one particular fractal, the Mandelbrot Set. The first example recalls Edna St. Vincent Millay's sonnet "Euclid alone has looked on Beauty bare" (discussed in Chapter 2). In "Fractals"[15] Diana Der-

[15] Der-Hovanessian includes the following footnote to explain the word "Fractal" in the title of her poem: "a geometry invented in 1975 by Benoit Mandelbrot to find order in chaotic shapes and processes."

Hovanessian has written a poetic tribute to Benoit Mandelbrot that begins with a direct reference to Millay's poem, through the use of an epigraph.

> *Euclid alone has looked on beauty bare*
> Edna St. Vincent Millay

Euclid alone began to formulate
the relation of circle, plane and sphere
in equations making it quite clear
it was symmetry that we must contemplate.

He left the jagged convoluted form,
rambling rivers, wind turbulence and rain;
ignored clouds, coastline and storms,
and the whorl of tree, skin and terrain,

to map the triangle, cone and square.
Euclid measured order and left the knot
of chaos to be unraveled by Mandelbrot
who found truth in the course of blood and air.
Euclid looked on beauty stark and bare,
but Mandelbrot appraised her everywhere.

(Der-Hovanessian 276)

One could learn a great deal about fractal geometry from this poem. Der-Hovanessian shows how Mandelbrot extended mathematics beyond Euclidean geometry as she describes many of the qualities associated with fractals ("jagged convoluted form" /"rambling rivers"/ "whorl of skin"). She also shows that fractals may be found virtually everywhere. "Fractals" is both a sonnet and a contribution to the tradition of writing about mathematical advances. Thus, its form replicates that of Millay's poem at the same time that its ideas expand those of Millay. In many ways, Der-Hovanessian's poem reinterprets both poetic and mathematical traditions. Speaking metaphorically, we could say that the poem is an *iteration* of those traditions.

While incorporating the ideas of fractal geometry into a poem about human relations, Robin Chapman uses words, rather than computer pixels, to create her own illustration of the Mandelbrot Set. To structure the poem, "Escaping the Mandelbrot Set," Chapman employs the basic properties of fractals — self-similarity over many scales and iteration. She also alludes to the specific properties of the Mandelbrot Set, using it as a powerful metaphor for escaping the "set" of one's experiences.

Def: Equation yielding a fractal pattern self-similar
at every level of magnification within a range of values.

She says
The coffee is fine
Though it could have been stronger
And cream would be nice.

She says
The weather today
Is, yes, fine, though cold
For summer and more rain likely tonight.

She says
The summer's going well,
Of course awfully fast and won't last
Long enough to get done what she'd planned.

She says
The marriage was ten good years
And then ten bad, and she's learned
A lot since, though of course it's lonely.

She says
Buying a new cappuccino maker,
Espresso roast, and best jam for her bread
Is frivolous, but we only has one life.

(Chapman 116)

Using the language of fractal geometry to discuss the poem, we can say that the repeating first line, "She says," opens a new iteration in each stanza. And with each iteration we see a different portion of the scene in a different scale. The stanzas move outward, from the coffee, to the weather; the summer and the marriage. But each observation is somehow qualified: the coffee *could have been* stronger; the weather is fine, *though* cold. The scene is observed over a variety of scales; each time the details of the scene are a bit rough or imperfect, while the overall similarity from scene to scene is certainly recognizable. Stanza by stanza, we see the woman's life in a larger context, but we simultaneously gain a deeper view into the inner qualities of that life. It is possible that this pattern could continue indefinitely, extending into further and further iterations and revelations, but instead of endlessly replicating the pattern, Chapman interrupts it.

The last stanza begins just as the others have, with the words "She says," but "she" suddenly takes the poem in a different direction and considers buying a new cappuccino maker. With this single deliberate and pleasure-seeking act, the woman in the poem is escaping the domestic pattern of her life, as well as the iterative pattern of the poem. This stanza reminds us of the point just beyond the edge of the Mandelbrot Set whose iteration "escapes" the set. The implied metaphor here is that the woman *is* that point, escaping the Mandelbrot Set, just as the title suggests. At the end of the poem, we return to the coffee that was first mentioned in the opening stanza. Now the coffee is recognizable in the approximate self-similarity of the cappuccino. By introducing the coffee again, in this new guise, Chapman uses the poetic possibilities of metaphor to help us imagine a new beginning for the woman.

Poetic Dimensions of Self-similarity

A poet's knowledge of fractal geometry may directly influence the word choice and internal structure of a poem, even if the poem isn't specifically *about* fractals. Judging from its title, "Fractal: Repetition of Form over a Variety of Scales," Pattiann Rogers obviously intends for the following poem to be understood within the context of fractal geometry.

This moment is a single blue jay,
a scramble of flint, sapphire iron,
spiking blue among the empty brambles
and vines wound like skeins back
upon themselves through the dun forest
of thistle spurs and thorns.

And this moment is as well the brambled
skeleton of the jay, anthracite spine,
thorny blades and femurs, tangle
of knuckled twigs flittering
through an equal flitter of jointed
sticks, fines and husks of wind.

And as well again, this split second
is the single blue-black pod of jay heart
thiddering among a bramble of rib bones
inside the tufts, the bristled
capsules of forest and winter barbed
and strung with dusk.

And the jay's call is this same
instant, a cry of release slivered
and shaped by the tangle of bones
and scrub woods, by the bolus wound
of winter air, thatched and spurred,
through which it travels.

And this moment is a single point
of sun wrapped and templing in the black
pathways of the blue jay's eye, like a heart
shuddering in a tangle of bones,
like a bird in a shifting knot of forest,
a call in a skeletal patch of winter,
winter in a weaving clutch of dusk, a moment
tangling within the string and bristle
of its own vocabulary.

God is a process, a raveled nexus
forever tangling into and around the changing
form of his own moment -- pulse and skein,
shifting mien, repeating cry
of loss and delivery.

(Rogers 55-56)

The concept of dimension is critical here. Just as the visible details and length of a shoreline increase when we magnify a photographic image, so too, can we see more as Rogers zooms in on the forest scene, focusing first on the blue jay, then on its skeleton, its heart, its call, its eye, and finally on the "raveled nexus" that Rogers equates to God. An iterative process helps organize the poem, and self-similarities appear in each stanza. At the end, our attention moves past the physical details of the scene to consider its spiritual dimensions.

Visual and aural elements blend to create another type of fractal patterning. This is accomplished with clusters of recurring synonymous words:

barbed, bristled, spurred
skein, tangle, strung, string, tangling
husk, pod, capsule
skeleton, femur, bones, spine.

Rogers' repetition of these word groups gives the poem a dense visual and tactile quality. Within what might appear to be a chaotic tumble of woods and weeds, there are interwoven patterns. Just as powerful, however, is the repetition of closely-related *sounds*. Many of these words are onomatopoeic, that is, they sound like the ideas they represent ("bristling," "flittering," or "shuddering"), so we also have a series of echoes, much like the sounds in a forest. This poem embodies many of the qualities the poet Alice Fulton described when crafting a definition for fractal poetry, where, according to Fulton, texture and recurring cluster words, as well tactile and dynamic elements are all critical.[16]

Reading with Fractals in Mind

In the contemporary poems presented in the preceding pages, the poets are aware of fractals and are consciously trying to bring the discoveries of science into their work. Just as fractal geometry has influenced the writing of poetry, it has also influenced how poetry is read and interpreted. As an analogy, it's helpful to remember that once the concept of fractals became accepted in contemporary mathematics, many scientists began to see previously-unrecognized fractal patterns in structures ranging from weather systems to the human body. Like the scientists who were willing to reconsider existing photographs, maps, or computer print-outs in light of fractal discoveries, we, too, may be able to observe new patterns if we apply our knowledge of fractals to reading poetry. In the pattern poem in Figure 4.46, for example, we can discern the visual properties of a now-familiar fractal.

When this anonymous eighteenth-century pattern poem and an iteration of the Hilbert Curve are placed alongside one another, the similarities in form stand out. Nonetheless, the intricate German poem, printed as a New Year's

---

[16] See Alice Fulton, "Fractal Amplifications: Writing in Three Dimensions" in *Feeling As a Foreign Language*. St. Paul, MN: Graywolf Press, 1999. Literary theory and criticism based on fractal science are drawing increasing interest, and several writers have done groundbreaking work in this area. Fulton has written extensively on what is meant by "fractal poetry," working toward a definition of the term, testing her definition against her own poetry, and applying this new aesthetic to the work of poets as diverse as Emily Dickinson and Ezra Pound. Fulton describes fractal poetry as resisting classification, existing between the structures of traditional, formal verse and the openness of contemporary free verse. She asserts that "as free verse broke the pentameter, fractal verse can break the poem plane or linguistic surface" (5). In this context, it is important to remember that Fulton is referring to fractal qualities found in the language and form of poetry; not to poems *about* fractals. Of "fractal poetry," Fulton says, "[. . .] digression, interruption, fragmentation, and lack of continuity will be regarded as formal functions, rather than lapses into formlessness" (58). Other writers, including Paul Lake, have examined the influence of fractal science on theories of poetics. See Lake's essay "The Shape of Poetry" and others on the subject in Kurt Brown's collection *The Measured Word: On Poetry and Science*. Athens, GA: University of Georgia, 2001.

greeting (Higgins 85), was composed and published a century before Peano and Hilbert discovered the "pathological" curves that revolutionized modern mathematics.

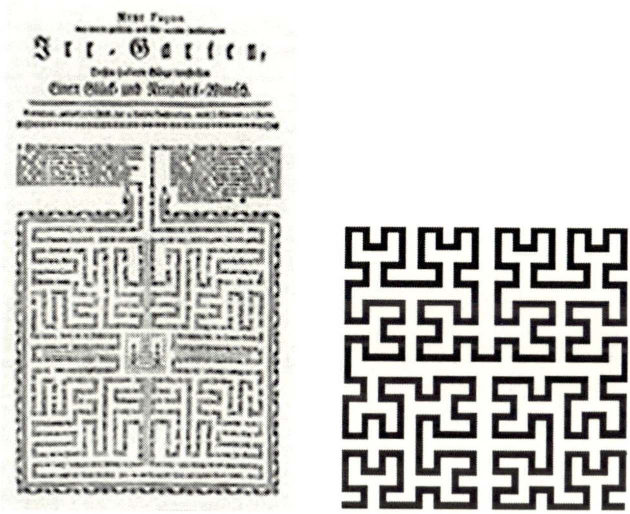

Figure 4.46

Probably pleased by the intricate symmetries of the pattern and challenged by the problem of creating a verbal maze, the poet unknowingly constructed a pattern poem whose visual properties are similar to those used in constructing Hilbert's space-filling curve.

Just as we may find resemblances to fractals in the physical form of poems that pre-date fractal geometry, we may also identify fractal qualities in the language of earlier poets.[17] What happens if a reader with some knowledge of fractals simply re-reads a favorite work keeping fractal principles in mind? If we look closely at William Shakespeare's "Sonnet 73," first introduced in Chapter 1, we find a pattern in the use of metaphor that suggests the fractal quality of iteration.

[17] Principles of fractal geometry have been employed in interesting, experimental ways as tools for literary analysis. The psychologist and critic Lucy Pollard-Gott, for example, has studied Wallace Stevens' poetry looking for fractal patterns in the recurrence of a single sound, a word, or a series of words with a shared root. Using a system of shaded and unshaded boxes, Pollard-Gott found a pattern she identifies as "fractal dust" in the recurrence of the related words *knowledge*, *known*, and *know* in Canto I of Stevens' poem "The Sail of Ulysses."

"Sonnet 73" illustrates well how the structure of the English sonnet may be used to develop and unify three separate ideas, in this instance, reflections on the impending loss of a loved one.

That time of year thou mayst in me behold
When yellow leaves, or none, or few, do hang
Upon those boughs which shake against the cold,
Bare ruined choirs where late the sweet birds sang.
In me thou seest the twilight of such day
As after sunset fadeth in the west,

Which by and by black night doth take away,
Death's second self that seals up all in rest.
In me thou seest the glowing of such fire
That on the ashes of his youth doth lie,
As the deathbed whereon it must expire,
Consumed with that which it was nourished by.
This thou perceiv'st, which makes thy love more strong
To love that well which thou must leave ere long.

(Shakespeare 1465)

As with any Shakespearean sonnet, there are many inherent patterns in this poem: patterns of meter, patterns of rhyme, and a division into three 4-line quatrains, followed by a 2-line couplet. In this sonnet, there is a pattern, too, in the arrangement of ideas. In each quatrain, the speaker compares himself to an element of nature and with each successive comparison, the focus draws in more closely, from a season, to an hour, to the last moments of the waning fire that heats and lights the speaker's room. As he constructs the series of comparisons, Shakespeare uses metaphors that reflect an organic progression. With each quatrain, the metaphor draws closer to the speaker himself, as death approaches. From a matter of months, to hours, to minutes, the time left to the speaker grows shorter each time it is measured. The inevitability implied in the comparisons becomes more and more immediate until the impending loss of the hearth's heat and light is all that remains. The repeated comparisons drawn between the speaker and various natural cycles create an iterative use of metaphor; furthermore, since the idea expressed in each quatrain is similar to the idea expressed by the poem overall, we could also say that, like a fractal image where each repetition approximates the original, here, the sonnet exhibits self-similarity over a variety of scales.

The Japanese poet Nanao Sakaki has spent years walking great distances, living outdoors, and writing with clarity and humor about the universe and his place in it. The subjects of Sakaki's poetry range from the minute and the mundane to the cosmic. His poem "A Love Letter" can be read as a fractal map of our location in the universe. It also serves as a poetic model for the fractal properties of approximate self-similarity, iteration, and dimension without making any overt reference to fractal science itself.

Within a circle of one meter
You sit, pray and sing.

Within a shelter ten meters large
You sleep well, rain sounds a lullaby.

Within a field a hundred meters large
raise rice and goats.

Within a valley a thousand meters large
Gather firewood, water, wild vegetables and Amanitas.

Within a forest ten kilometers large
Play with raccoons, hawks,
Poison snakes and butterflies.

Mountainous country Shinano
A hundred kilometers large
Where someone lives leisurely, they say.

Within a circle one thousand kilometers large
Go to see the southern coral reef in summer
Or winter drifting ices in the sea of Okhotsk.

Within a circle ten thousand kilometers large
Walking somewhere on the earth.

Within a circle one hundred thousand kilometers large
Swimming in the sea of shooting stars.

Within a circle one million kilometers large
Upon the spaced-out yellow mustard blossoms
The moon in the east, the sun west.

Within a circle ten billion kilometers large
Pop far out of the solar system mandala.

Within a circle ten thousand light years large
The Galaxy full blooming in spring.

Within a circle one million light years large
Andromeda is melting away into snowing cherry flowers.

Now within a circle ten billion light years large
All thoughts of time, space are burnt away
There again you sit, pray and sing.
You sit, pray and sing.

(Sakaki 12-13)

This poem does not mention fractals in its title, nor does it resemble a fractal in appearance. However, the fractal-sensitive reader will find reading it a pleasure. "A Love Letter" uses strong visual imagery as it invites us to step further and further back from the first actions of the poem until we are ten billion light years away, past the point where Andromeda has melted into "snowing cherry flowers." As we look for fractal elements, we can see that Sakaki has used a simple repeating phrase ("within the ...") to tether the stanzas of the poem together while the subject is placed within larger and larger concentric circles in the universe. The widening circles and repeated phrases develop the self-similarity, as the location of the subject "you" is described in greater and greater distance from the original one meter circle. However, the self-similarity is only approximate since in some iterations "you" seems to disappear, replaced by an ambiguous subject or by "the galaxy" or "Andromeda." The scale increases gradually and in recognizable measurements, from one meter to ten meters, and eventually to ten billion light years. And the iterations are mapped by the progression of stanzas; with each step outward, the action becomes more fantastic, more cosmic, until the single figure "you" is finally the focus again, sitting in the center of the same circle, now ten billion light years large, praying and singing, as if prepared to begin the process all over again. This pattern of spinning out into an elaborate iterative map is reminiscent of the repeated magnifications of the antennae of the Mandelbrot Set, where each successive magnification reveals another, new version of the entire set. In Sakaki's poem, however, it is language, not computer graphics, that brings the image to life and creates its structure.

Fractals As a Link Between Mathematics and Poetry

Fractals may be the most complex and the most subtle example of patterns found in both mathematics and poetry. Mathematics provided the formulas from which the first fractal images were created, and it is those computer-generated images that non-scientists most often think of when they hear the word "fractals." When poets borrowed ideas from fractal geometry and applied them to the reading and writing of poetry, they made a remarkable intellectual leap. The historical influences of science on art are numerous, but when we look closely how the interest in fractals spread beyond the scientific community, we have a vivid, contemporary example of science influencing art within the period of a few years. But that influence does not operate in one direction alone. By bringing fractal geometry into their work, poets have also helped shape the layperson's understanding of this new field of mathematics. Fractals serve as one of the most provocative bridges between mathematics and poetry. When fractal concepts, as understood and applied in the two disciplines, are discussed in tandem, countless intellectual and creative possibilities emerge for both the poet and mathematician.

Finally, for individuals who are intrigued simply by the challenge of finding things that are hidden, learning about fractals makes it possible to understand in a small way the pleasure of discovery described by Benoit Mandelbrot:

> The nature of fractals is meant to be gradually discovered by the reader, not revealed in a flash by the author.
>
> And the art can be enjoyed for itself.
>
> (*The Fractal Geometry of Nature* 5)

Chapter 5 – Patterns for the Mind

5.1 The Mathematical Mind: Proof, Paradox, and Infinity

This chapter is about patterns for the mind — patterns of abstract thinking that possess strong aesthetic qualities and challenge our intellect. The patterns are more conceptual than the concrete patterns found in numbers, form, and shape. They are patterns we use to express ideas about truth and contradiction. We also use them when we write or speak about the mysteries of our world, the profundities and complexities of what we believe. Welcome to patterns for the mind!

Mathematical Proof

To begin, imagine you are creating something new. You want it to be perfect and beautiful, elegant and flowing. You are familiar with similar work preceding your effort, but you don't want to be limited by the vision of others. While form, content, and structure are important, creativity is crucial. The right opening may make all the difference, but it is the conclusion that will have lasting significance.

Just what is it you are creating? While it could be an essay, a work of art, or a poem, your hypothetical creation could also be a mathematical proof. Most people would accept the idea that poetry has aesthetic and creative qualities, but fewer would associate beauty, elegance, and vision with mathematics.

In order to understand why mathematicians value the aesthetics of a proof, we should understand that they strive to write in a way that is not only clear, but also elegant and thought-provoking. They commonly describe a classic mathematical work as beautiful, inventive, lively, exciting, witty, or bold – perhaps even dramatic, passionate, mysterious, or thrilling. These are hardly adjectives the average person would use in describing a math class or textbook. It may surprise you that in describing their discipline, mathematicians sometimes compare it to poetry. In *Men of Mathematics*, E. T. Bell quotes the nineteenth-century mathematician Karl Weierstrass as saying, "A mathematician who is not also something of a poet will never be a complete mathematician" (Bell xix). More recently, in a 2004 National Academy of Science interview, Ingrid Daubechies, the Princeton mathematician who popularized wavelet analysis, creates an analogy between writing a poem and writing a proof. She recalls meeting the poet Richard Kenney when they were both named MacArthur Fellows and describes discovering similarities in their creative processes as follows:

> [. . .] When he tried to explain to me the experience of writing a poem it sounded very much like I would describe the experience of trying to find the proof or theorem I feel is true [. . .] He has a hunch, he feels there is something, and so on, and he has to tease the words out. And I have to tease the mathematical proof. I have to find the exact route that will lead me to that proof and he has to find the exact words. It is a very similar emotional experience for a very different end [. . .] (Daubechies)

With this background, it is not surprising to find aesthetics important in writing mathematical proof. But what are the essential characteristics of a well-constructed proof? A mathematician begins with spare, uncluttered steps that proceed logically. Perhaps the proof relies on very few assumptions, or it takes existing assumptions and twists them in an unexpected way. Possibly the proof can be generalized to a variety of other situations and will result in further insights, or it may connect two areas of mathematics that previously appeared unrelated. Some proofs have repercussions far beyond mathematics, such as the proof that $\sqrt{2}$ is an irrational number, which led to the end of the philosophy of rationalism during the Pythagorean era of ancient Greek mathematics.

What exactly are mathematicians trying to prove? Mathematics is based on a system of axioms — principles that are self-evident without needing proof. Axioms are the starting assumptions from which other statements are logically derived. Any statement beyond the basic axioms must be proven before it can be accepted. Such a statement is called a *theorem*, from the Greek root "thea" which means the act of seeing. Other mathematical terms associated with proof are *lemma*, which is an auxiliary theorem proved beforehand and then used in the proof of the main theorem, and *corollary*, which is a theorem that is a consequence of the proof of the main theorem. Lemmas may be thought of as "mini-theorems" that are the stepping-stones leading to a main theorem, while corollaries are the "mini-theorems" that are simple, direct consequences of the main theorem.

In some ways, writing a proof is similar to writing a poem. Both require creative energy and a willingness to challenge known ideas. Poets and mathematicians often arrive at an epistemological moment when they move past what is known and familiar, bringing innovation and sometimes controversy to their respective fields. On a practical level, both poems and proofs usually require rewriting or refinement to "tighten the language" or make them more succinct.

Like other mathematical milestones, famous proofs have also been celebrated in poetry, both in ancient and modern times. When the Royal Astronomical Society in 1919 accepted Einstein's proof of his theory of relativity, the British astronomer Sir Arthur Eddington commemorated the occasion with the following poem:

The Clock no question makes of Fasts and Slows,
But steadily and with a constant Rate it goes.
And Lo! The clouds are parting and the Sun
A crescent glimmering on the screen – It shows! –
It shows!

Five minutes, not a moment left to waste,
Five minutes, for the picture to be traced –
The stars are shining, and coronal light
Streams for the Orb of Darkness – Oh make haste!

For in and out, above, about, below
"Tis nothing, but a Magic Shadow-show
Played in a Box whose candle is the Sun
Round which we Phantom Figures come and go.

Oh leave the Wise our measurements to collate.
One thing at least is certain, LIGHT has WEIGHT.
One thing is certain and the rest debate –
Light-rays , when near the Sun, DO NOT GO STRAIGHT.

(Eddington 146)

Here's a more contemporary example of using poetry to commemorate a momentous mathematical proof. For the last three hundred and fifty years, the greatest "brain teaser" of mathematics has been Fermat's Last Theorem. Pierre de Fermat was a seventeenth-century mathematician who studied equations of the form $x^n + y^n = z^n$, called Diophantine Equations after the ancient Greek algebraist Diophantus. Fermat's Theorem stated that $x^n + y^n = z^n$ had no integer solutions for $n > 2$ except for the trivial solution $x = y = z = 0$. On the other hand, when $n = 2$, the Diophantine Equation is something quite familiar, namely the Pythagorean Theorem, $x^2 + y^2 = z^2$, which has many integer solutions, such as:

$$3^2 + 4^2 = 5^2$$

$$5^2 + 12^2 = 13^2$$

$$8^2 + 15^2 = 17^2$$

Fermat could find no integer solutions to the equation $x^3 + y^3 = z^3$, but more importantly, he claimed that he had a marvelous general proof that no solutions existed for any equation of the form $x^n + y^n = z^n$ when $n > 2$. He wrote a hasty note about the proof in the margin of his copy of a text by Diophantus, but stated he just could not fit the proof itself in the narrow margin of the page!

For over three hundred years mathematicians sought such a proof. The theorem did not seem that difficult to prove, and yet time and again the mathematical community came up short. The modern day English mathematician, Andrew Wiles, was intrigued by the problem as a teenager and had tried tackling it both as a layman and again at various points in his mathematical career. Finally, in 1993, after seven years of intense work, he claimed to have proven the theorem, but there was an error in his argument that kept people on the edge of their mathematical seats. Would his proof hold up? Could he fix the error in his analysis? Would this be one more time that Fermat stumped the mathematical community? It would be another year, involving collaboration with Richard Taylor, a fellow Cambridge University mathematician, before Wiles' proof of Fermat's Theorem was declared complete. It is a long proof, running about one hundred and fifty pages, that utilizes twentieth-century mathematical ideas unknown to Fermat. So, in one sense the mystery is solved. Fermat was, indeed, correct in his statement of the theorem, but mystery still surrounds the theorem. Did Fermat actually have a proof of his theorem, and if so, can that proof be rediscovered? We may never be able to answer these questions.

As with Einstein's Theory of Relativity, mathematicians celebrated Wile's achievement in verse. In 1995, at the conclusion of a ten-day mathematics conference in which Wiles and Taylor presented their proof, the mathematician Jeremy Teitelbaum issued a poetic challenge. He asked the audience to contribute poems, both serious and fanciful, about the solving of Fermat's Last Theorem. The submissions ranged from limericks to multi-stanza poems and were published in true late twentieth-century format on Dr. Teitelbaum's web page. The two examples below are from Barry Mazur and

Marion Cohen, mathematicians who may be familiar to the reader for their more conventional publications[18].

Sing Fermat

When Fermat Vapours clog our loaded Brows
With furrow'd Frowns,
When stupid downcast Eyes th'external Symptoms
Of some
Gap
Within our Proof express,
Or when in sullen Dumps
With Head incumbent on expanded Palm,
Moping we sit,
Our Gaulloise snuffed, deform'd,
Sing then, Oh Wiles, and Taylor-Wiles!
Oh trio: put Fermata to our Toils.

(Mazur)

Fermat's Last Theorem Proven

Fermat said the proof was too large
to fit in the right or left marg-.
True, back of the paper
or proof made to taper
might help, but he said, "I'm in charge".

Now, Wiles didn't mind paper waste.
In fact, it was true to his taste
to use up whole reams
to realize his dreams
and he crossed out instead of erased.

---

[18] Barry Mazur holds the position of Gerhard Gade University Professor at Harvard University and is the author of *Imagining Numbers (particularly the square root of minus fifteen).* Marion Cohen is a poet and mathematician who teaches at the University of the Sciences in Philadelphia and at the University of Pennsylvania. Her most recent publication is *Crossing the Equal Sign.*

Fermat was all snickers and smiles
as he smugly stayed clear of the aisles
and he thought "they'll be glum
"but that proof will succumb
"though it's going to take quite a-Wiles."

(Cohen)

Like the patterns of numbers, form, and shape we have discussed in earlier chapters, patterns of proof can be categorized in a variety of ways. Some of the best-known types of proof include:

Direct Proof - Using axioms, definitions, and other known theorems, the statements flow logically to a conclusion.

Proof by Induction - In order to show a statement holds for all Natural Numbers, an initial case is proven to be true. An nth or general case is assumed to be true; finally, if the assumption from the general case is true, then the $n+1^{st}$ case can be proven.

Proof by Contradiction - The statement of the theorem is assumed not to be true. When a logical contradiction occurs under this assumption, the statement of the theorem is required to be true.

Proof by Exhaustion - The statement of the proof is divided into a finite number of cases, each of which is proven separately.

A simple example of a direct proof is shown below:

Theorem: The sum of two odd numbers larger than 1 is even.
Proof:
Let m and n be Natural Numbers (i.e. 1, 2, 3, . . .)
Then $2m + 1$ and $2n + 1$ are two odd numbers (even number plus one)
$(2m + 1) + (2n + 1) = 2m + 2n + 2 = 2(m + n + 1)$
$2(m + n + 1)$ is even since it is a multiple of 2.

Many of us remember well the proofs of high school geometry. For some this was pure joy – seeing the pattern of geometric figures as a puzzle that was fun to solve. For others, geometric proof was a form of torture designed specifically to prevent high school graduation. Mathematical proof is such an emotional trigger that mathematicians, as well as former math students, have developed humorous categories of proof. Some of the authors' favorites include:

Proof by Intimidation - What? You don't see that this is obvious???

Proof by Hand Waving and Dismissal - This is so trivial as to not need further clarification.

Proof by Imagination - Pretend for a moment that this is true. . .

Proof by Assignment - Because we are pressed for time, I'll leave this proof as a homework exercise.

Proof by Pedagogical Limitation - The proof of this theorem is well beyond the scope of this class. . .

Proof by Sameness - This proof is really identical to the last proof, so we'll omit it.

Proof by Cumbersome Notation - This proof would require several alphabets, many symbols, and more boards than we have in the classroom.

Proof by Inaccessibility - The proof was known to the 8th century mathematicians of China, but the reference has been lost.

Closely related to proof is the concept of fallacy. In mathematical circles, a fallacy is a statement that is known to be false, such as 5 = 0, that may be "proven" through a series of steps containing an error most people will miss. Let's "prove" that 5 is indeed equal to 0.

Theorem: 5 = 0

Proof:

1. Let $a = 3$ and $b = 2$
2. $a + b = 5$
3. Assume $a + b = c$
4. $(a + b)(a + b) = c(a + b)$
5. $a^2 + 2\,ab + b^2 = ca + cb$
6. $a^2 + ab - ca = c$

The fallacy occurs in line 7 when we divide both sides of the equation by the term $(a + b - c)$ which is equivalent to zero. Since one can never divide by zero, we introduced fallacious reasoning at that point. Fallacious proof may appear to be both true and false when it is first presented. In fact, the proof contains a step made in error — an error that is not obvious. We could say this theorem and its proof are actually an example of a paradox.

Paradox

In mathematics there are three types of paradox:

> An absurd proposition that arises from fallacious reasoning, such as the proof in which we showed 5 = 0.
>
> A theorem that is true, but is so strange or incredible that our intuition tells us it must be false, such as the statement that the Koch Snowflake has finite area, but infinite perimeter (as described in Chapter 4).
>
> Logical paradox that requires creative argument, such as Zeno's Paradox (first described in Chapter 1) claiming motion is impossible since there will always be half the distance remaining as we try to reach our destination.

Some paradoxes have shaken the foundations of mathematics, such as the "pathological curves" of Hilbert and Peano, described in Chapter 4, that fill a two dimensional space with a one-dimensional object. To resolve other paradoxes, mathematical thinking required a new methodology, such as the development of infinite series used to explain Zeno's paradox. Still other paradoxes remain unresolved.

Let's look at a few examples of paradox. First, there is the mathematician's "old standard" that 0 = 1. This paradox is based on fallacious reasoning, as in the proof that 5 = 0, but it does not involve division by zero.

Theorem: 0 = 1

Proof:

$0 = 1 - 1$

$0 = (1 - 1) + (1 - 1) + (1 - 1) + (1 - 1) + (1 - 1) + \dots$

Now rearrange the terms on the left side of the equation.

$0 = 1 + (-1 + 1) + (-1 + 1) + (-1 + 1) + (-1 + 1) + \dots$

$0 = 1 + 0 + 0 + 0 + 0 + \dots.$

$0 = 1$

The fallacy comes from rearranging the terms of an infinite sum as if it were a finite sum. By leaving the first 1 alone and assuming that all the subsequent terms will cancel out, we have made the fallacious assumption that the series ends.

One of the most famous paradoxes is the Barber's Paradox, proposed by Bertrand Russell, a twentieth-century British logician and philosopher. The statement of the paradox is straightforward: a male barber claims to shave every man who does not shave himself and only those men. But who shaves the barber? It can't be the barber because he only shaves men who do not shave themselves. But if he is a man who does not shave himself, then by the original statement he must shave himself. Thus we have an impossible situation arising from the barber's claim. This logical paradox had a profound effect on mathematics and led to new developments in the area of Set Theory. Mathematicians had to find new ways of thinking about and defining sets in order to resolve the apparent contradiction.

A final example is the paradoxical inn known as Hilbert's Hotel, presented by David Hilbert, the German mathematician who lived from 1862 to 1943 and whose work on the Hilbert Curve is described in Chapter 4. Hilbert's Hotel has an infinite number of rooms and each room is occupied. A new guest arrives and, although the hotel is full, the desk clerk promises to find an empty room. This is accomplished by moving the occupant of room n to room $n+1$, thus freeing up room 1 for the new guest. Of course, to settle this one new arrival we must move every guest already staying in the hotel. The occupant of room 1 moves to room 2, the occupant of room 2 moves to room 3, etc., as seen in Figure 5.1.

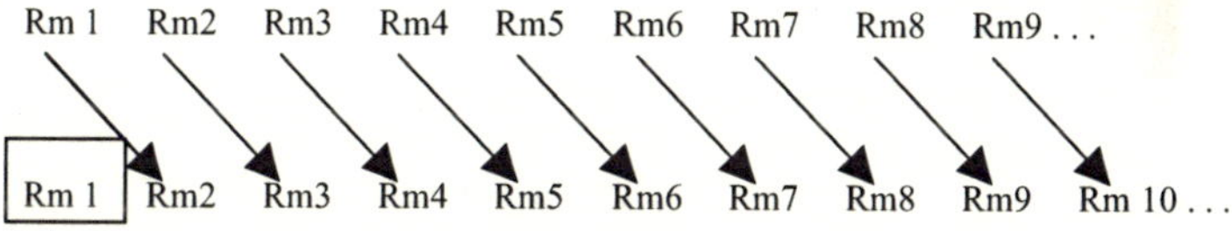

Figure 5.1

Just as the guests are settling in for the night, a new dilemma arises – a bus arrives at Hilbert's Hotel with an infinite number of passengers. What is the desk clerk to do? With a little thinking he decides to move each guest in room n to room $2n$. Thus the guest in room 1 moves to room 2, the guest in room 2 moves to room 4, the guest in room 3 moves to room 6, etc. Now all of the even-numbered rooms are occupied, but all of the odd-numbered rooms are vacant to accommodate the bus passengers. A visual illustration of this solution is shown in Figure 5.2.

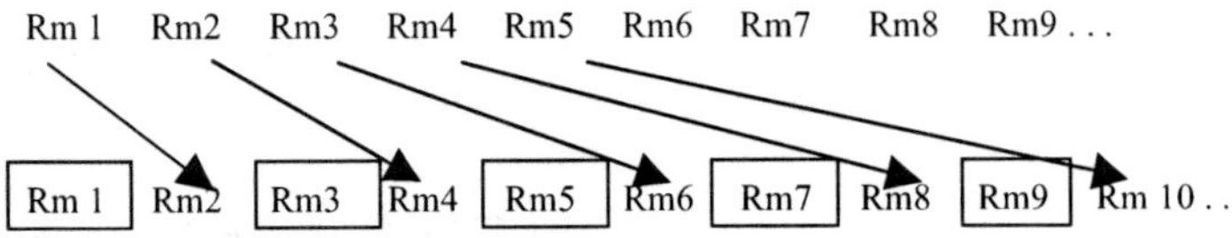

Figure 5.2

The tale continues on, recounting how Hilbert's Hotel can even accommodate an infinite number of buses, each with an infinite number of passengers. Hilbert's Hotel is relying upon the ideas of countable infinity that were presented in Chapter 1, where the subset of even counting numbers is exactly the same size the entire set of all counting numbers. What if infinity were uncountable?

Infinity Revisited

In Chapter 1, we explored the idea of Georg Cantor's countable infinity. Cantor's results were paradoxical to the mathematicians of his time and, in fact, were rejected by many prominent colleagues. Cantor used the symbol $\aleph_0$ to represent the size or cardinality of countably infinite sets, such as the Integers and the Rational Numbers. In this section we return to Cantor's work, this time looking at uncountable infinity. An uncountably infinite set, such as the Real Numbers, is one that cannot be placed in correspondence with the Natural Numbers. Cantor referred to the size or cardinality of such sets as *C*, for Continuum. While $\aleph_0$ represents the cardinality of a countably infinite set, *C* represents an infinite set that has larger cardinality than $\aleph_0$. Cantor's work revealed an infinite hierarchy of ever-increasing infinities.

Cantor's formal proof that the Real Numbers are not countably infinite is complex, but may be summarized informally through the technique of Proof by Contradiction given in the boxed text below. In addition to the Real Numbers, other infinite sets of size *C* include:

The set of Irrational Numbers
The set of Real Numbers between 0 and 1
The set of all points on a straight line
The set of all points in a plane
The set of all points in three-dimensional space.

Theorem: The Real Numbers are uncountably infinite
Proof:
Assume that the Real Numbers are countably infinite.
Each Real Numbers has an infinite decimal representation:

Real Number		Decimal Approximation
1	=	1.00000. . .
$\sqrt{2}$	=	1.41421. . .
7/3	=	2.33333. . .
...		...

Then, the entire set of Real Numbers can be placed in a one-to-one correspondence with the Natural Numbers, as shown in the list below. Every Real Number must appear on the right side of the list!

Natural Numbers	Real Numbers
1	1.000...
2	1.41421. . .
3	2.3333333. . .
4	65.0169332. . .
5	0.331869412. . .
6	100.543087. . .
. . .	. . .

If the Real Numbers were countably infinite, then all Real Numbers would appear in this list. You may create a new Real Number that is NOT in the list by choosing any number for the integer digit and creating the decimal portion as follows:
Choose the first decimal place digit to be different from the first decimal place digit of the first Real Number in the list. Choose the second decimal place digit so that it is different from the second decimal place digit of the second number in the list, and so on. This is illustrated below by underlining the changing numbers.

1.<u>0</u>00. . .	1st decimal place digit must not be 0
1.4<u>1</u>421. . .	2nd decimal place digit must not be 1
2.33<u>3</u>3333. . .	3rd decimal place digit must not be 3
65.016<u>9</u>332. . .	4th decimal place digit must not be 9
0.3318<u>6</u>9412. . .	5th decimal place digit must not be 6
100.54308<u>7</u>. . .	6th decimal place digit must not be 7
. . .	. . . .

One such new number is 5.234512. . . This number is not on the list since it is different from each number on the list. Thus we've contradicted the original statement that every real number is in our countably infinite list.

If you find Cantor's ideas about uncountable infinity difficult to grasp, you are not alone. Although Cantor's new way of thinking about the infinite ultimately revolutionized both mathematics and philosophy, initially his work met staunch resistance from some of his contemporaries.

Why does infinity evoke such strong reactions? The infinite is not only a mathematical concept, but is also tied to deeply-held philosophical and theological beliefs. Infinity has no physical manifestation – it can only be imagined. Poetry may be a perfect medium for presenting ideas of infinity. In these opening lines of William Blake's "Auguries of Innocence," the profound shifts in scale easily capture our imagination.

> To see a world in a Grain of Sand
> And a Heaven in a Wild Flower,
> Hold Infinity in the palm of your hand,
> And Eternity in an hour. . .
>
> (Blake 481)

In each line, a single object or moment contains the potential for something far greater than itself. Using an imaginary magnifying glass, Blake makes it possible for us to visualize an entire world in a grain of sand, then we raise our eyes to see endless fields of flowers within a single wild flower. When he speaks of infinity and eternity, he shifts the focus to humankind, where Blake's lines suggest the unlimited possibilities contained in every human being, in every hour.

5.2 Ideas that "tease us out of thought"

William Blake is by no means alone in using poetry to consider the infinite. In fact, all three of the concepts covered in this chapter, proof, paradox, and infinity, have fascinated poets as well as mathematicians. Often difficult to comprehend, these concepts engage our curiosity, fuel our imaginations, and make us think critically about claims of truthfulness and the meaning of boundlessness. However, the ways these concepts are employed in poetry are very different from their use in mathematics. The poet is usually not tying to establish what is meant by proof, paradox, or infinity. Instead, these ideas are important in the poet's work because of what they have come to stand for in our imagination.

The statements in the following scenarios may help illustrate these differences:

> Statement 1 - Pretend for a moment you are sitting in a coffee shop where you overhear one person whisper passionately to another, "I will prove how much I love you right now."
>
> Statement 2 - An injured athlete watches from the sidelines as his team wins a close victory. He says to himself, "I'll remember this moment longer than all of them."
>
> Statement 3 - Or, after explaining the concept of multiple infinities, a mathematics professor waves a piece of paper in the air, claiming, "I can easily capture infinity on this very page."

What are some of the observations we can make about these statements?

> On first reading each statement seems untrue, or at least very unlikely.
>
> Each statement makes a claim that would need to be supported or explained in order to be convincing.
>
> Each statement automatically provokes further inquiry, inviting the listener to ask "how" or "why."

What is also interesting, though perhaps not immediately clear, is that the answer to the problem raised in each statement could best be explained by using language that is non-literal, language that makes comparisons or offers illustrations. As we examine how poets address the "mind-stretching ideas" of proof, paradox, and infinity in their work, we will pay special attention to the use of language. Eventually, we'll encounter our three opening statements again, re-cast in poetry, but first we need to review what it is that makes the poet's use of language different from that of the mathematician.

Considering Language

In poetry, words are chosen and arranged as carefully as the numbers and symbols of mathematics, yet sometimes, despite our ability to read and decode many other kinds of writing, we may find poetry as difficult to understand as the most complicated math problem. And indeed, a poem and a math problem cannot be "solved" in the same way. The mathematician's unambiguous language of series, proofs, or logarithms is of little help when reading a poem. Why is that? For one thing, the words the poet uses are shaded with meaning. Words we understand in one context may appear in another, only to be used quite differently. The poet's intention in choosing words is also important: some words may be used for their subtle or allusive qualities; other may be chosen because they are explicit, jarring, or even nonsensical. The poet may also borrow words from other languages or invent new words, known literally as *neologisms*.

> There are countless names for the specialized uses of language in poetry. Here are some simple definitions for a few of the most common literary terms:
>
> Symbols - This term refers to material objects that a writer uses to stand for other things. As the writer Henry James said, "symbols cast long shadows." Virtually any object — from flowers, to rings, to fire, a river, or the moon — may cast shadows that create layers of meaning in a poem.
>
> Irony - This identifies an indirect form of expression where words don't mean what they say. The contradictions, and sometimes absurdities, in ironic language are often used for powerful effects in social commentary and satire. One familiar form of irony is sarcasm, when words mean the opposite of what is said or written, as in "I love it when you put your feet on the table."

Figurative language - This is a collective term for "figures of speech." Two of the most familiar figures of speech are metaphor and simile; both are created by drawing comparisons between dissimilar things. Something that is difficult to explain may be described in terms of something else that is more familiar, or the comparison may be drawn to show qualities not obvious at first. At the heart of the metaphor or simile, there are often important shared qualities that give the figure of speech its depth. (Even Robert Burns' well-known metaphor "My luve is a red, red rose," embodies more than a greeting card sentiment, implying, "My love is fragile, beautiful, but sometimes thorny or blighted.") In a metaphor, one thing is described as if it were another, while in a simile, a connecting word, such as "like" or "as" is used to make the connection.

Figuring out the Figurative

For an illustration of the poet's language at work, let's look closely at John Keats' poem "Ode on a Grecian Urn." We will see how Keats layers the figurative language to take us well beyond the literal meanings of words. In the opening lines, Keats uses several metaphors, which we show in italics, to describe an ancient urn that has survived, intact, through many centuries:

"Thou *still unravished bride* of quietness,
    Thou *foster child of silence and slow time*,
Sylvan historian, who canst thus express
    A flowery tale more sweetly than our rhyme:"

Each of these metaphors compares the urn and the story that wraps around its form to a person with whom the reader is already familiar (bride, foster child, historian). By creating this series of comparisons, Keats helps the reader see the urn as more than a piece of pottery. When he refers to the urn as if it were a human being with a specific identity or role, Keats substitutes *known* qualities for things we don't know (the urn's origins, the significance of the tale it tells). In doing so, he modifies those human identities to make them unique to the urn. The *bride* is married to quietness; the *foster child* is reared by silence and slow time. In this way, Keats gives us a bridge into thinking about the urn in a non-literal way, but he also offers much to ponder as we try to grasp the complexity in the metaphors. As the poem goes on, Keats' reflection on the urn leads to increasingly abstract ideas until the final lines of the poem, when he writes:

> "Thou, silent form, dost tease us out of thought
> As doth eternity: Cold Pastoral!
> When old age shall this generation waste,
> That shalt remain, in midst of other woe
> Than ours, a friend to man, to whom thou say'st,
> Beauty is truth, truth beauty -- that is all
> Ye know on earth, and all ye need to know."
>
> (Keats 252-53)

The experience he describes, that of being "teased out of thought" is familiar to most of us. When we daydream about an object in a museum or wonder about faces in an old photograph, we can easily leave our present-day concerns behind and allow our thinking to drift off. We may also experience spontaneous, imaginative thinking when we stare into the night sky and wonder about the past, the future, even the possibility of life on other planets. At times like this, we let go of our familiar surroundings and routines to follow the intellectual equivalent of a mysterious and unfamiliar path in the woods.

Sometimes, such reveries are inspired by things far less tangible than artifacts, photographs, or stars. Just as mysterious physical objects do, poetry and mathematics may lead us into ways of thinking we have never experienced before. When we have an "aha!" realization because we are beginning to understand the concept of multiple infinities, or when our curiosity is piqued by a poet's enigmatic use of language, as in the lines, "Beauty is truth, truth beauty -- that is all/Ye know on earth, and all ye need to know," we have moved beyond the limit of what is known and familiar. In "Ode on a Grecian Urn," Keats is describing a quality of thought we have all experienced, when, as curious creatures, we are challenged to ponder the imponderable.

Where Lies the Proof?

If we return to the concept of proof, we'll find that poetic "proof" often relies heavily on figurative language. In Chapter 1, we introduced Elizabeth Barrett Browning's most famous sonnet, in which the speaker begins by asking, "How do I love thee?" She then answers her own question, saying, "Let me count the ways." If we pay close attention to the lines where Browning proves her love by counting, measuring, and comparing its qualities to other things, known and imagined, we can follow the figurative language of metaphor. Within the sonnet's 14 lines, Browning devises a method, similar to a direct proof, by which she can bring together a universe of detail, from the spiritual to the mundane, to demonstrate the extent of her love by comparing it to other known and accepted experiences. She offers the proof necessary to support hypothetical Statement 1 presented at the beginning of this section: "I will prove how much I love you right now."

How do I love thee? Let me count the ways.
I love thee to the depth and breadth and height
My soul can reach, when feeling out of sight
For the ends of Being and ideal Grace.
I love thee to the level of everyday's
Most quiet need, by sun and candle light.
I love thee freely, as men strive for Right;
I love thee purely, as they turn from Praise.
I love thee with the passion put to use
In my old griefs, and with my childhood's faith.
I love thee with a love I seemed to lose
With my lost saints, -- I love thee with the breath,
Smiles, tears, of all my life! -- and, if God choose,
I shall but love thee better after death.

(Browning 43)

Instead of devising a system of proof to develop the poem's ideas, a poet may refer to proof in a more abstract way. In the following lines, taken from "Reflections on Poetry + Proof," by Marga Rose Hancock, the speaker asks rhetorically, "where lies the proof of what we all believe?"

"Does beauty heal?" "Does order give us peace?"
Where lies the proof of what we all believe?
Art embraces science without cease
As by design, our ideals interweave.

(Hancock)

The answer itself is as thought-provoking as Keats' urn since it suggests that the proof of our beliefs lies in the constant blending of art and science, that "beauty" [art] and "order" [science] have the power to improve our lives and our world. The "interweaving" of art and science promoted in these lines might make you think of the application of the Golden Ratio to art, architecture, even to the composition of an ideal human figure. Or you may be reminded of how spirals and labyrinths have served as a guide to meditation, or how symmetry is used in the design of a cathedral. Can you think of other instances, large or small, where art embraces science to provide healing or peace?

In the poem “When I Heard the Learn'd Astronomer,” Walt Whitman explores the symbolic implications of proof, associating the astronomer’s proof, figures, charts, and diagrams with a tiring, enervating lecture.

When I heard the learn'd astronomer,
When the proofs, the figures, were ranged in
  columns before me,
When I was shown the charts and diagrams, to add, divide,
  and measure them,
When I sitting heard the astronomer where he lectured
  with much applause in the lecture-room,
How soon unaccountable I became tired and sick,
Till rising and gliding out I wander'd off by myself,
In the mystical moist night-air, and from time to time,
Look’d up in perfect silence at the stars.

(Whitman 309)

Perhaps the mathematical statements and drawings are meant to represent an institutionalized or academic approach to learning that forgets the beauty of the very thing it purports to study, in this case, the stars in the night sky. When the speaker goes outside alone to look at the stars, he describes his experience in language very different from that used to describe the lecture; the air is now “mystical,” and the silence is “perfect.” The fatigue and tedium symbolized by proofs are left far behind.

Poetry and Paradox

The concept of paradox underlies the central idea of many poems. How do poets use paradox? Let’s begin with a definition of the word as it is used in a literary context.

In *A Dictionary of Literary Terms*, J. A. Cuddin defines paradox as “an apparently self-contradictory (even absurd) statement which, on closer inspection, is found to contain a truth reconciling the conflicting opposites” (479-80). This definition is helpful when we read the following poem by Emily Dickinson.

Success is counted sweetest
By those who ne'er succeed.
To comprehend a nectar
Requires the sorest need.

Not one of all the purple host
Who took the flag today
Can tell the definition,
So clear, of victory

As he, defeated, dying,
On whose forbidden ear
The distant strains of triumph
Break, agonized and clear.

(Dickinson 7)

Beginning with the provocative claim that "success is counted sweetest/ by those who ne'er succeed," Dickinson deftly proves the truth in that apparent paradox by helping us imagine the longing and the despair of the man who didn't succeed, the person who was left behind, defeated and dying. The celebration of those in the victorious "purple host" is portrayed as mob-like and ignorant, while only the dying man can fully understand what he has lost. Thus Statement 2, spoken by the sidelined athlete, becomes clearer as we understand the paradox in his words, "I'll remember this moment longer than all of them."

At first glance, the title of Stephen Crane's poem "War is Kind," seems to be a perfect example of Cuddin's definition of "an apparently self-contradictory (even absurd) statement." However, when we study the poem closely, we realize that it does not fulfill the second part of the definition, that the contradictory statement, "is found to contain a truth reconciling the conflicting opposites." There is no truth hidden in this poem; the conflicting opposites in the statement "war is kind" are not reconciled, but are made painfully clear in the poem's details.

Do not weep, maiden, for war is kind.
Because your lover threw wild hands toward the sky
And the affrighted steed ran on alone,
Do not weep.
War is kind.

Hoarse, booming drums of the regiment,
Little souls who thirst for fight,

These men were born to drill and die.
The unexplained glory flies above them,
Great is the Battle-God, great, and his Kingdom --
A field where a thousand corpses lie.

Do not weep, babe, for war is kind.
Because your father tumbled in the yellow trenches,
Raged at his breast, gulped and died,
Do not weep.
War is kind.

Swift blazing flag of the regiment,
Eagle with crest of red and gold,
These men were born to drill and die.
Point for them the virtue of slaughter,
Make plain to them the excellence of killing
And a field where a thousand corpses lie.

Mother whose heart hung humble as a button
On the bright splendid shroud of your son,
Do not weep.
War is kind.

(Crane 81-82)

War is NOT kind, as Crane graphically illustrates with the dying lover, the thousand corpses of those born to drill and die, the fatherless son, and bereaved mother. Rather than presenting a paradox, Crane's poem is instead an example of verbal irony, where the speaker means something quite different from what he or she says, no matter how well supported the statement may appear to be.

Knowing about mathematical paradoxes will prepare you well for the figurative language in the poem "Depression" by Wendy Cope. As you read, think about the ways Cope uses words from science and math to describe the condition of depression. The phrases "snail-like," "narrowing spiral," and "Zeno's arrow" all allude to ideas we have discussed earlier, but if you were not familiar with them, the poem might be much more difficult to read and its meaning much harder to grasp.

You lie, snail-like, on your stomach -
I dare not speak or touch,
Knowing too well the ways of our kind-
The retreat, the narrowing spiral.

We are both convinced it is impossible
To close the distance.
I can no more cross this room
Than Zeno's arrow.

(Cope 298)

Once you understand the allusions Cope is making, her poem becomes clearer and opens up to interpretation. Through her use of simile ("snail-like"), we can visualize the curled posture of the person she's addressing, as well as the isolated, solitary shell-building that separates the two people. The reference to Zeno's paradox and the inability of an arrow to traverse the distance between two points is a painful metaphor for the inability to bridge the distance that may separate people, whether physically or psychologically. Cope's use of figurative language based on mathematical concepts makes this poem moving and memorable.

Poetic Concepts of the Infinite

As we've seen, the mathematical concepts of proof and paradox may be used poetically in many ways. Proof may be applied to a poem's structure, or the poet may employ paradox to force us to examine the veracity of an idea. When poets turn their attention to infinity, usually it is not because of the structural or rhetorical possibilities it offers, since it would be difficult to create an infinitely-long poem or a poem whose ideas are literally infinite. Instead, poets are drawn to the idea of the boundlessness of infinity, but how can a writer put into words a concept that poses such a challenge to human understanding?

Often, poets describe infinity by comparing it to things that are more familiar, such as the uncountable grains of sand on a beach or the stars in the sky. In these lines by Jack Kerouac, the moon suggests infinity:

The moon her magic be, big sad face
Of infinity...

(Kerouac 111)

And for Rabindranath Tagore, the sky is seen as infinite:

On the seashore of endless worlds children meet.
The infinite sky is motionless overhead...

(Tagore 41)

Other poetic references to infinity may be drawn from religious texts, myths, works of art, or scientific treatises. Or the language used to conceptualize infinity may be borrowed from technology, from information provided by mechanical devices such as telescopes, cameras, or computers.[19]

Infinity may, in turn, be used to describe other unquantifiable aspects of human experience, such as desire:

> [. . .] This is the monstrosity in love, lady, that the will is
> infinite, and the execution confined; that the desire is
> boundless, and the act a slave to limit.
>
> (Shakespeare, *Troilus and Cressida*, III: ii, 74-77)

or suffering:

> [. . .] Suffering is permanent, obscure and dark,
> And shares the nature of infinity.
>
> (Wordsworth, "The Borderers" III. 1543-44)

In the following poems, infinity and the infinitesimal are represented in vastly different ways: as a solitary fall into oblivion, as an unimaginable point between two realities, and as a metaphor for the creative act.

In "The Infinite," a poem first published in 1819, Giacomo Leopardi writes of an individual lost in a "sea" of infinity, where his thinking drowns, and he is tempted by the allure of being shipwrecked. Here, as in many other poems, the concept of infinity leads easily to that of eternity.

> Always dear to me was this lonely hill,
> And this hedge, which from so great a part
> Of the farthest horizon excludes my gaze.
> But as I sit and watch, I invent in my mind
> endless spaces beyond, and superhuman
> silences, and profoundest quiet;
> wherefore my heart
> almost loses itself in fear. And as I hear the wind
> rustle through these plants, I compare

[19] For those interested in further reading on this subject, the scholarship of Marjorie Hope Nicolson offers insight into the influence of science on literature. In *Science and Imagination*, she writes of the influence of the telescope and the microscope on the imagination, citing many examples of English poetry. In *Mountain Gloom and Mountain Glory*, Nicolson examines the "Aesthetics of the Infinite," as poetry moved from the 17th to 19th centuries and ideas of infinity and eternity were transferred from "a God of Power and a God of Benignantly to Space, and then to the grandeur and majesty of earth" (393).

that infinite silence to this voice:
and I recall to mind eternity,
And the dead seasons, and the one present
And alive, and the sound of it. So in this
Immensity my thinking drowns:
And to shipwreck is sweet for me in this sea.

(Leopardi 105)

The landscape over which the speaker is gazing suggests the vastness of infinity, but it also evokes another feeling, the isolation and alienation of the individual. To lose himself in this infinity offers a kind of solace, as he imagines drowning, shipwrecked in infinity. By combining the "sweetness" of an escape into oblivion with the terror we associate with a shipwreck, Leopardi creates for the reader a powerful, visceral impression of the infinite.

In contrast to Leopardi's evocation of the infinite through the visual image of a grand vista and the terrifying occurrence of a shipwreck, Ilse Bing writes of the infinitesimal as a compelling point toward which we are drawn. Rather than a physical place, it is more a moment in time. Bing's description of that point begins in a mathematical context:

infinitesimal is the nearest to zero
infinitesimal is so small
that it is no longer something
but it is not yet nothing

if jumping into the water
you detect the instant
when you are no more in the air
and not yet in the water
you grasp the infinitesimal

this infinitesimal instant
lies at the point
where the possible and the impossible
touch each other

(Bing 12)

Michael L. Johnson writes about infinity by focusing on another work of art, the M.C. Escher design "Circle Limit III" (seen in Figure 3.36). Using familiar mathematical terminology, he begins by describing how Escher creates the illusion of infinity. He then focuses on the design itself (much as

Keats' poem focused on the details of the Grecian urn), before stepping back to consider the artist and his work in a broader perspective:

M.C. Escher's *Circle Limit III*

This tessellated hyperbolic plane
is definitely non-Euclidean,
though inside, not on, its circumference
points correspond. Outside is emptiness.

The fish swim back and forth but cannot sense
how they progressively grow small or large
by distance from the unreachable edge
where hypercycles shrink to nothingness.

Thus miracled infinity is viewed
but only by a god of finitude.

(Johnson "M.C. Escher" 282)

The poem, like Escher's drawing, creates its own illusory world, beyond which there is emptiness. Of course, in the end, both the work of art and the poem are limited by the very real constraints or "finitude" of paper and ink. Although it requires a "god of finitude," that is, Escher himself, to help us see infinity, the infinity-suggesting design is still bounded by the limits of the surface on which it is drawn. Nonetheless, it shows that we can indeed capture infinity on a piece of paper, just as the lecturer claims in hypothetical Statement 3.

This poem makes us realize the difficulties any artist faces when trying to create an image of infinity. For Johnson, the concept of infinity, while miraculous, can be also used to draw our attention to the act of creation. The artist and the poet capture the infinite on the page, using whatever illusions are necessary, to make visible or give words to those things we can barely imagine.

Chapter 6 – Conclusion

As you've seen, there are fairly obvious similarities between some patterns in mathematics and poetry (think of magic squares and word square poems), as well as similarities that are much more subtle (the mathematical patterns in closed form poems). There are also links based on intellect and imagination (poems written on fractals or infinity).

Some mathematicians write verse and some poets are good problem solvers, but there have been scholars across history who have excelled in both fields, either as authors of great poetry and excellent mathematical treatises, or as practitioners in one field and writers about the other field. Some whose work we studied while researching this book include Eratosthenes of Greece (276-194 B.C.E.); Omar Khayyám (1048-1131 C.E.); Blaise Pascal (1623-1662); Lewis Carroll, the pen name of Charles Lutwidge Dodgson, (1832-1898); Piet Hein (1905-1996); and Jacob Bronowski (1908-1974). These thinkers found some parallel patterns in math and poetry, but more importantly, were able to bridge differences of thought in the two subjects.

Despite many instances where mathematics and poetry overlap in structure or idea, there are also enormous differences between the two disciplines, differences we would not attempt to bridge or minimize. Because of the complexities of proof, paradox, and infinity, we brought these concepts together in Chapter 5 to highlight one of the most important differences between our fields, that is, the contrast between the concrete methodology of mathematics and the figurative nature of poetry. The works of mathematicians and poets are fundamentally different, as are the ways we approach and study those works. Nonetheless, the intellectual and creative affinities between the two disciplines are compelling. The mathematician and the poet are both drawn toward the mysterious, continually searching for ways to stretch our understanding of the world. Both find joy in the creative act of their work, and both are passionate about the importance of that work, as an expression and embodiment of human curiosity, intellect, and beliefs.

We hope your own curiosity has been stirred by the discussions, the poems, the mathematical analysis, and the illustrations you have seen here. We also hope you will continue to look for patterns in the physical world and in the world of ideas, whether those patterns are based on numbers or composed with words, captured in a photograph, or exhibited in the faces in a crowd, the outline of buildings against a skyline.

In patterns, ideas take shape. They are given life, recorded, and passed on to others. When we explore the patterns in mathematics and poetry, we sit alongside the creative thinkers and artists who came before us. We follow their computations, study their drawings, read and listen to their words, and share their emotions, attracted always by something "deep in the structure," as described in "This Poem" by Barbara Jordan:

> Let the form be a garden in wild wilderness,
> a hyacinth language, a turning in wind
>   when marginal influences
> disrupt the flow.
>
> Build thought as a bee does,
> one concern at a time, a hexagonal symmetry
>   deep in the structure [. . .]

(Jordan 55)

Permissions Acknowledgements

All reasonable steps have been taken to contact copyright holders of material used in this book. The authors would be pleased to make suitable arrangements with any whom they have not been able to reach. Copyright holders may contact the authors through the publisher.

Aisenberg, Nadya. "Measures," *Measures*. Cliffs of Moher, Co. Clare, Ireland: Salmon Publishing, 2001. Used by permission of the publisher.

Apollinaire, Guillaume. "Il Pleut / It's Raining" by Guillaume H. Apollinaire, Translated by Roger Shattuck, from *Selected Writings*, copyright © 1971 by Roger Shattuck. Reprinted by permission of New Directions Publishing Corp.

Bishop, Elizabeth. "A Miracle for Breakfast" from *The Complete Poems 1927-1979* by Elizabeth Bishop. Copyright © 1979, 1983 by Alice Helen Methfessel. Reprinted by permission of Farrar, Straus and Giroux, LLC.

Chapman, Robin. "Escaping the Mandelbrot Set" from *Images of a Complex World: the Art and Poetry of Chaos* by Robin Chapman and Julien Clinton Sprott. Singapore: World Scientific Publishing Co. Pte. Ltd.: 2005. Used by permission of Robin Chapman.

Cherry, Kelly. "Sequence Sonnet" from *Relativity: A Point of View,* by Kelly Cherry (Louisiana State University Press, 1977; Carnegie Mellon University Press, 2000). Used by permission of the author.

Cohen, Marion. "Fermat's Last Theorem Proven." Reprinted with permission from Marion Cohen — first appeared on Jeremy Teitelbaum's Fermat's Last Theorem Poetry Challenge web page.

Cope, Wendy. "Depression," *The Faber Book of 20th Century Women's Poetry*, ed. Fleur Adcock. London: Faber and Faber, Ltd., 1987. Used by permission of the author and the PFD Group Ltd.

Cummings, E.E. "lis//-ten" and ")when what hugs stopping earth than silence is" are reprinted from *Complete Poems 1904-1962*, by E.E. Cummings, edited by George J. Firmage, by permission of W.W. Norton & Company. Copyright © 1991 by the Trustees for the E.E. Cummings Trust and George James Firmage.

Der-Hovanessian, Diana. "Fractals," *Verse & Universe: Poems about Science and Mathematics*, ed. Kurt Brown. Minneapolis, MN: Milkweed Editions, 1998. First appeared in *Songs of Bread, Songs of Salt* published by Ashod

Press in 1990. Copyright © Diana Der-Hovanessian. Reprinted by permission of the author.

Escher M. C. "Circle Limit III" © 2007 The M. C. Escher Company — Holland. All rights reserved. Reprinted with permission of The M. C. Escher Company, www.mcescher.com.

Escher M. C. "Symmetry Drawing E63" © 2007 The M. C. Escher Company — Holland. All rights reserved. Reprinted with permission of The M. C. Escher Company, www.mcescher.com.

Garrison, Denis M. "Nautilus" used by permission of Denis M. Garrison.

Hancock, Marga Rose. "Reflections on Poetry + Proof" used by permission of the author and American Institute of Architects.

Holmes, Janet. "The Pattern" from *The Physicist at the Mall.* Tallahassee, FL: Anhinga Press, 1994. Used by permission of the publisher.

Hughes, Langston. "The Negro Speaks of Rivers," copyright © 1994 by The Estate of Langston Hughes, from *The Collected Poems of Langston Hughes* by Langston Hughes, edited by Arnold Rampersad with David Roessel, Associate Editor. Used by permission of Alfred A. Knopf, a division of Random House, Inc.

Johnson, Michael L. "Fibonacci Time Lines" is reprinted from *The Unicorn Captured* (Lawrence, KS: Cottonwood Review Press, 1980) with permission from the author, Michael L. Johnson. "M.C. Escher's Circle Limit III" is reprinted from *Familiar Stranger* (Lawrence, KS: Flowerpot Mountain Press, 1983) with permission from the author, Michael L. Johnson.

Jordan, Barbara. "This Poem" from *Trace Elements* by Barbara Jordan, copyright © 1998 by Barbara Jordan. Used by permission of Penguin, a division of Penguin Group (USA) Inc.

Mazur, Barry. "Sing Fermat" published by permission of the author.

North, Charles. "Sonnet." Reprinted from *Cadenza* © 2007 by Charles North, by permission of Hanging Loose Press.

Oliver, Mary. "Something," from *Why I Wake Early* by Mary Oliver. Copyright © 2004 by Mary Oliver. Reprinted by permission of Beacon Press, Boston, MA.

Quiller-Couch, Arthur T. "A New Ballad of Sir Patrick Spens" was originally published by J.M. Dent & Sons Publishers and The Orion Publishing Group. All attempts at tracing the copyright holder were unsuccessful.

Rogers, Pattiann. "Fractal: Repetition of Form over a Variety of Scales." Pattiann Rogers, *Eating Bread and Honey* (Minneapolis: Milkweed Editions, 1997). Copyright © 1997 by Pattiann Rogers. Reprinted with permission from Milkweed Editions.

Sakaki, Nanao. "A Love Letter." From Sakaki, Nanao. *Break the Mirror*. Nobleboro, ME: Blackberry Books, 1996. Used by permission of the publisher.

Sandburg, Carl. "Arithmetic" in *The Complete Poems of Carl Sandburg*, copyright © 1950 by Carl Sandburg and renewed 1978 by Margaret Sandburg, Helga Sandburg Crile and Janet Sandburg, reprinted by permission of Harcourt, Inc.

Solt, Mary Ellen. "Moon shot" and "FORSYTHIA" used by permission of the author.

Wiltz, Chris. "Icebreaker" used by permission of the author.

Bibliography

Aisenberg, Nadya. "Measures," *Measures*. Cliffs of Moher, Co. Clare, Ireland: Salmon Publishing, 2001.

Apollinaire, Guillaume. "Il Pleut," *Selected Writings of Guillaume Apollinaire*, trans. Roger Shattuck. New York: New Directions, 1971.

Aros Fractals software. Issaquah, WA: Aros Magic Research, 1998, <http://www.arosmagic.com/Fractals> [accessed 25 September 2007] (File Download Area).

Bell, E. T. *Men of Mathematics*. New York: Simon and Schuster, Inc., 1937.

Bing, Ilse. "Infinitesimal," *Against Infinity: An Anthology of Contemporary Mathematical Poetry*, eds. Ernest Robson and Jet Wimp. Parker Ford, PA: Primary Press, 1979.

Bishop, Elizabeth. "A Miracle for Breakfast," *The Complete Poems*. New York: Farrar, Straus, and Giroux, 1969.

Blake, William. "Auguries of Innocence," *The Poetry and Prose of William Blake*, ed. David V. Erdman. Garden City, NY: Doubleday & Co., Inc., 1965.

Bohn, Willard. *The Aesthetics of Visual Poetry 1914 - 1928*. Cambridge: Cambridge University Press, 1986.

Boulton, Marjorie. *The Anatomy of Poetry*. London: Routledge & Kegan Paul Ltd., 1953.

Briggs, John. *Fractals The Patterns of Chaos: Discovering a New Aesthetic of Art, Science, and Nature*. New York: Simon and Schuster, 1992.

Brown, Kurt, ed. *The Measured Word: On Poetry and Science*. Athens and London: The University of Georgia Press, 2001.

_______. *Verse & Universe: Poems about Science and Mathematics*. Minneapolis: Milkweed Editions, 1998.

Browning, Elizabeth Barrett. "Sonnet XLIII," *Sonnets from the Portuguese* reprinted from *Poems by Elizabeth Browning*, 4th ed. Chapman and Hall, 1856; New York: Heritage Club, 1938.

Buchanan, Scott. *Poetry and Mathematics*. New York: The John Day Company, 1929; reprinted Philadelphia and New York: J. B. Lippincott Co., 1962.

Chapman, Robin. "Escaping the Mandelbrot Set," Chapman and Julien Clinton Sprott. *Images of a Complex World: The Art and Poetry of Chaos*. Singapore: World Scientific Publishing Co. Pte. Ltd., 2005.

Charles, Dorthi. "Concrete Cat," *Literature: An Introduction to Fiction, Poetry, and Drama*. ed. X.J. Kennedy, 2nd ed., Boston: Little, Brown, 1979.

Cherry, Kelly. "Sequence Sonnet," *Relativity: A Point of View*. Pittsburgh, PA: Carnegie Mellon University Press, 2000.

Christensen, Inger. *Alphabet*, trans. Susanna Nied. New York: New Directions Publishing, 2001.

Cohen, Marion. "Fermat's Last Theorem Proven," *Permission to Add: Math Teaching Limer's pdf.* <http://marioncohen.com/downloads.cfm> [accessed 25 September 2007] (page 48 of 48).

Cook, Theodore Andrea. *The Curves of Life*. New York: Dover Publications, 1979.

Cope, Wendy. "Depression," *The Faber Book of 20th Century Women's Poetry*, ed. Fleur Adcock. London: Faber and Faber, Ltd., 1987.

Crane, Stephen. "War is Kind," *The Poems of Stephen Crane*, ed. Joseph Katz. New York: Cooper Square Publishers, Inc., 1966.

Cuddin, J. A. ed. *A Dictionary of Literary Terms,* revised ed. New York: Penguin Books, Doubleday & Co., Inc., 1980.

Cummings, E.E. "lis/-ten," *AnOther E.E. Cummings*, ed. Richard Kostelanetz. New York: W. W., Norton & Co., 1998.

_______. ")when what hugs stopping earth than silent is" from *50 Poems*, in *Poems 1923-1954*, New York: Harcourt, Brace and Co., 1926.

Daubechies, Ingrid. "InterViews" National Academy of Sciences, April 2004. <http://www.nasonline.org/site/PageServer?pagename= INTERVIEWS_Ingrid_Daubechies> [accessed 18 July 2007] (track 3).

de Beaulieu, Eustorg. "Gloire à Dieu seul," in Hélène Harvitt, *Eustorg de Beaulieu: A Disciple of Marot.* New York: Columbia University Press, 1918; New York: AMS Press, 1966.

Dekker, Thomas. "The Honest Whore," *Fantasia Mathematica*, ed. Clifton Fadiman. New York: Simon & Schuster, 1958; New York: Copernicus, 1997.

Der-Hovanessian, Diana. "Fractals," *Verse & Universe: Poems about Science and Mathematics*, ed. Kurt Brown. Minneapolis, MN: Milkweed Editions, 1998.

Devaney, Robert L. *Chaos, Fractals, and Dynamics: Computer Experiments in Mathematics*. Menlo Park, CA: Addison-Wesley Publishing Co., 1990.

Dickinson, Emily. "Success is Counted Sweetest," *Final Harvest: Emily Dickinson's Poems*. Boston: Little, Brown & Co., 1961.

Doplicher, Fabio. "Asymmetry of the Universe," *New Italian Poets*, eds. Dana Gioia and Michael Palma. Brownsville, OR: Story Line Press, 1991.

Duckworth, George E. *Structural Patterns and Proportion in Vergil's Aeneid: A Study in Mathematical Composition*. Ann Arbor: The University of Michigan Press, 1962.

Eddington, Arthur. "The Clock no question makes of Fasts and Slows," in Amir D. Aczel, *God's Equation: Einstein, Relativity, and the Expanding Universe*. New York: Random House, 2000.

Edwards, A.W.F. *Pascal's Arithmetic Triangle*. Baltimore: Johns Hopkins University Press, 2002.

Fadiman, Clifton, ed. *Fantasia Mathematica*. New York: Simon & Schuster, 1958; New York: Copernicus, 1997.

Finch, Annie and Kathrine Varnes, eds. *An Exaltation of Forms: Contemporary Poets Celebrate the Diversity of Their Art*. Ann Arbor: University of Michigan Press, 2002.

Fischler, Roger. "How to Find the 'Golden Number' Without Really Trying," *The Fibonacci Quarterly*, 19 (1981): 406 - 410.

Flegg, Graham. *Numbers: Their History and Meaning*. New York: Schocken Books, 1983.

Frucht, William, ed. *Imaginary Numbers*. New York: John Wiley & Sons, Inc., 1999.

Fulton, Alice. *Feeling as a Foreign Language*. St. Paul, MN; Graywolf Press, 1999.

Gardner, Martin. *Mathematical Circus*. Washington, DC: Mathematical Association of America, 1992.

Garland, Trudi Hammel. *Fascinating Fibonaccis: Mystery and Magic in Numbers*. Parsippany, NJ: Dale Seymour Publications, 1987.

Garrison, Denis. "Nautilus," *Nautiluses and Fibonaccis* (unpublished document provided by author). First published in <http://www.summit.net/writers_hood/02jul/tp06_07_02.htm> [accessed 14 August 2003].

Gomringer, Eugen. [Untitled Poem-Figure 5], *Concrete Poetry: A World View*, ed. Mary Ellen Solt. Bloomington: Indiana University Press, 1970.

Hancock, Marga Rose. "Reflections on Poetry + Proof," *AIArchitect*. <http://www.aia.org/aiarchitect/thisweek03/tw0516/0516margaverse.htm> [accessed 25 September 2007].

Herbert, George. "Easter Wings," *The Works of George Herbert*, ed. F.E Hutchinson. Oxford: Oxford University Press, 1941.

Higgins, Dick. *Pattern Poetry: Guide to an Unknown Literature*. Albany, NY: State University of New York Press, 1987.

Holmes, Janet. "The Pattern," *Verse & Universe: Poems about Science and Mathematics*, ed. Kurt Brown. Minneapolis, MN: Milkweed Editions, 1998.

Holmes, Oliver Wendell. "The Chambered Nautilus," *The Complete Poetical Works of Oliver Wendell Holmes*. Boston: Houghton, Mifflin, and Co., 1895.

Holyoak, Keith J. and Paul Thagard. *Mental Leaps: Analogy in Creative Thought*. Cambridge, MA: MIT Press, 1995.

Hughes, Langston. "The Negro Speaks of Rivers," *The Collected Poems of Langston Hughes*, ed. Arnold Rampersad. New York: Alfred A. Knopf, 1994.

Huntley, H. E. *The Divine Proportion: A Study in Mathematical Beauty*. New York: Dover Publications, Inc., 1970.

Johnson, James Weldon. "Sence You Went Away," *101 Great American Poems*, ed. The American Poetry & Literacy Project. Mineola, NY: Dover Publications, Inc., 1998.

Johnson, Michael L. "Fibonacci Time Lines," *Verse & Universe: Poems about Science and Mathematics*, ed. Kurt Brown. Minneapolis: Milkweed Editions, 1998; first published in *The Unicorn Captured*. Lawrence, KS: Cottonwood Review Press, 1980.

_______. "M. C. Escher's *Circle Limit III*," *Verse & Universe: Poems about Science and Mathematics*, ed. Kurt Brown. Minneapolis: Milkweed Editions, 1998; first published in *Familiar Stranger*. Lawrence, KS: Flowerpot Mountain Press, 1983.

Jordan, Barbara. "This Poem," *Trace Elements*. New York: Penguin Books, 1998.

Julia's Dream software. Reinoud Lamberts, The Netherlands, 1991, <http://hypertextbook.com/chaos/92b.shtml> [accessed 10 May 2005] (Mandelbrot and Julia Sets).

Kaufman, Gerald Lynton. "Sigh-Angles" and "Spiralay," *Geo-metric Verse: Poetry Forms in Mathematics Written Mostly for Fanatics*. New York: The Beechhurst Press, 1948.

Keats, John. "Ode on a Grecian Urn," *The Selected Poetry of Keats*, ed. Paul de Man. New York: Signet Classic Poetry Series, 1966.

Kerouac, Jack. "The Moon," *Pomes All Sizes*. San Francisco, CA: City Lights Books, 1992.

Koshy, Thomas. *Fibonacci and Lucas Numbers with Applications*. New York: John Wiley & Sons, Inc., 2001.

Kostelanetz, Richard, ed. *Imaged Words & Worded Images*. New York: Outerbridge & Dienstfrey, 1970.

Knott, Ron. *Fibonacci Numbers and the Golden Section*. Surrey University, <http://www.mcs.surrey.ac.uk/Personal/R.Knott/Fibonacci/fib.html> [accessed 19 October 2006].

Lear, Edward. "There was an Old Man of the Coast," *The Complete Nonsense Book*. New York: Dodd, Mead & Co., 1959.

Leopardi, Giacomo. "The Infinite," *Introduction to Italian Poetry*, ed. Luciano Rebay. New York: Dover Publications, Inc., 1969.

Lesmoir-Gordon, Nigel, Will Rood, and Ralph Edney. *Introducing Fractal Geometry*. Lanham, MD: Totem Books (National Book Networks, Inc.), 2001.

Livio, Mario. *The Golden Ratio: The Story of Phi, the World's Most Astonishing Number*. New York: Broadway Books, 2002.

MacGillavry, Caroline H.. *Fantasy & Symmetry: The Periodic Drawings of M. C. Escher*. New York: Harry N. Abrams, 1976.

Mandelbrot, Benoit B. *Fractals: Form, Chance, and Dimension*. San Francisco: W. H. Freeman & Co., 1977.

_______. *The Fractal Geometry of Nature*. New York: W. H. Freeman & Co., 1983.

Maor, Eli. *To Infinity and Beyond: A Cultural History of the Infinite*. Princeton, NJ: Princeton University Press, 1991.

Mazur, Barry. "Sing Fermat," (unpublished document provided by author). First published in *Fermat's Last Theorem Poetry Challenge*, ed. Jeremy Teitelbaum. <http://www2.math.uic.edu/~jeremy/poetry.html> [accessed April 2005].

McKay, Claude. "If We Must Die," *The Passion of Claude McKay: Selected Prose and Poetry 1912 - 1948*, ed. Wayne Cooper. New York: Schocken Books, 1973.

Millay, Edna St. Vincent. "Euclid alone has looked on Beauty bare," *The Harp-Weaver and Other Poems*. New York: Harper & Brothers, 1920.

________. "Recuerdo," *A Few Figs from Thistles: Poems and Sonnets*. New York: Harper & Brothers, 1922.

________. "Small Hands Relinquish All," *Collected Poems*, ed. Norma Millay. New York: Harper & Row, 1945.

Mother Goose. Illustrated by Gyo Fujikawa. New York: Grosset & Dunlap, 1968.

Newcomb, Lawrence. *Newcomb's Wildflower Guide*. Boston: Little, Brown & Co., 1977.

Newman, James R. *The World of Mathematics: A Small Library of the Literature of Mathematics from A'h-mose' the Scribe to Albert Einstein*. Redmond, WA: Tempus Books of Microsoft Press, 1988.

Nicolson, Marjorie Hope. *Mountain Gloom and Mountain Glory: The Development of the Aesthetics of the Infinite*. Cornell University, 1959; reprinted as a Weyerhaeuser Environmental Classic, Seattle, WA and London: University of Washington Press, 1997

_______. *Science and Imagination*. Ithaca, NY: Cornell University Press, 1956.

North, Charles. "Sonnet," *Best American Poetry 2002*, ed. Robert Creeley. New York: Scribner Poetry, 2002.

Oliver, Mary. *Rules for the Dance: A Handbook for Writing and Reading Metrical Verse*. New York: Houghton Mifflin Co., 1998.

_______. "Something," *Why I Wake Early*. Boston, MA: Beacon Press, 2004.

The Oxford Annotated Bible with the Apocrypha (Revised Standard Version) eds. Herbert G. May and Bruce M. Metzger. New York: Oxford University Press, 1965.

Padgett, Ron, ed. *The Teachers & Writers Handbook of Poetic Forms, 2nd ed.* New York: Teachers & Writers Collaborative, 2000.

Peitgen, Heinz-Otto and P. H. Richter. *The Beauty of Fractals: Images of Complex Dynamical Systems*. New York & Berlin: Springer-Verlag, 1986.

Petrarch, Francesco. "Zephyr returns, and scatters everywhere," *Lyric Poetry of the Italian Renaissance*, ed. L.R. Lind. New Haven: Yale University Press, 1954.

Pinsky, Robert. *The Sounds of Poetry: A Brief Guide*. New York: Farrar, Straus, and Giroux, 1998.

Plotz, Helen, ed. *Imagination's Other Place: Poems of Science and Mathematics*. New York: Thomas Y. Crowell Co., 1955.

Pollard-Gott, Lucy. "Fractal Repetition Structure in the Poetry of Wallace Stevens," *Language and Style*, 19:3 (Summer 1986) 233-249.

Puttenham, George. "Her majestie, for many parts in her," *The Arte of English Poesie* (1589), ed. G. D. Willcock and A. Walker. Cambridge: Cambridge University Press, 1936; reprinted 1970.

Quiller-Couch, Arthur T. "A New Ballad of Sir Patrick Spens," *Fantasia Mathematica*, ed. Clifton Fadiman. New York: Copernicus, 1997. Originally published New York: Simon & Schuster, 1958.

Robinson, Edwin Arlington. "The House on the Hill," *Collected Poems*. New York: MacMillan Co., 1954.

Rogers, Pattiann. "Fractal: Repetition of Form over a Variety of Scales," *Eating Bread and Honey*. Minneapolis MN: Milkweed Editions, 1997.

Sakaki, Nanao. "A Love Letter," *Break the Mirror*. Nobleboro, ME: Blackberry Books, 1996.

Sandburg, Carl. "Arithmetic," *Complete Poems*. New York: Harcourt, Brace and Co., 1950.

Schattschneider, Doris. *M. C. Escher: Visions of Symmetry*. New York: Harry N. Abrams, 2004.

Shakespeare, William. "Sonnet 73," *The Complete Pelican Shakespeare*, gen. ed. Alfred Harbage. Baltimore, MD: Penguin Books, 1969.

_______. *The History of Troilus and Cressida, The Complete Pelican Shakespeare.*

Shattuck, Roger. "It's Raining," (trans.), *Selected Writings of Guillaume Apollinaire*, trans. Roger Shattuck. New York: New Directions, 1971.

Silliman, Ron. *Tjanting*. Cambridge, UK: Salt Publishing, 2002.

Siqueira, Rodrigo. "The Cantor Dust – Fractal Poetry," formerly accessed at: www.insite.com.br/rodrigo/misc/fracal/fractal_poetry.html

"Sir Patrick Spens," *Traditional British Ballads*. ed. Bartlett Jere Whiting. New York: Appleton-Century-Crofts, New Century Classics, 1955.

Solt, Mary Ellen. "Moon shot" and "FORSYTHIA," *Concrete Poetry: A World View*. Bloomington: Indiana University Press, 1970.

Strand, Mark and Eavan Boland. *The Making of a Poem: A Norton Anthology of Poetic Forms*. New York: W. W. Norton & Co., 2000.

Stevens, Peter S. *Handbook of Regular Patterns: An Introduction to Symmetry in Two Dimensions*. Cambridge, MA: The MIT Press, 1980.

Swinburne, Algernon Charles. "The Complaint of Lisa," *The Complete Poems of Charles Algernon Swinburne*, Vol. 3. New York: Harper & Brothers Publishers, 1904.

Tagore, Rabindranath. "On the Seashore," *The Collected Poems and Plays of Rabindranath Tagore*. New York: The Macmillan Co., 1913.

Thierer, John W. (revising author), William A. Niering and Nancy C. Olmstead (original authors). *National Audubon Society Field Guide to North American Wildflowers: Eastern Region,* rev. ed. New York: Chanticleer Press, Inc., 2001.

Thompson, D'Arcy. *On Growth and Form*. Cambridge, UK: Cambridge University Press, 1961.

Teitelbaum, Jeremy. *Fermat's Last Theorem Poetry Challenge*. <http://www2.math.uic.edu/~jeremy/poetry.html> [accessed April 2005].

Whitman, Walt, "When I Heard the Learn'd Astronomer," *Leaves of Grass*. New York: The Modern Library, 1921.

Wildman, Eugene, ed. *Anthology of Concretism*, 2nd ed. Chicago: Swallow Press, 1969.

Williams, Emmett, ed. *An Anthology of Concrete Poetry*. New York: Something Else Press, 1967.

Wilson, Robert M. "Fibonacci Memory," *I See Goddesses*. Allahabad, India: Cyberwit.net, 2004.

Wiltz, Chris. "Icebreaker," unpublished poem. Used with permission of author.

Wordsworth, William. "The Borderers," *The Poetical Works of William Wordsworth*, Vol. I. London: Oxford University Press, 1940.

Index

A

accentual verse, 32-33
- accentual-syllabic verse, 33-34

Aisenberg, Nadya, 65
- "Measures," 65

al-Karaji, 45
$\aleph_0$, aleph naught, 25,178
analogy, 5, 9-10, 129, 170
- analogical, 10

"Analogy, Mathematics, and Poetry" ("Patterns in Poetry and Mathematics") college course, 5
anapest, 34-36
Apollinaire, Guillaume, 100-101
- calligrammes, 100
- "Il Pleut," 100-101,127
- translation by Roger Shattuck, 101

assonance, 38

B

ballad, 39-41, 68
- "Sir Patrick Spens" traditional ballad and parody, 39-41

Barber's Paradox, 177
Bell, E. T., 169
- *Men of Mathematics,* 169

Bernoulli, Jacob (Jacques), 26
- *Ars Conjectandi,* 26

Bing, Ilse, 191
- "Infinitesimal," 191

Bishop, Elizabeth, 66-68
- "A Miracle for Breakfast," 66-69

Bishop, Morris, translator of Petrarch, 76
Blake, William, 180-181
- "Auguries of Innocence," 180

Boulton, Marjorie, 34-35
Bronowski, Jacob, 193
Brown, Kurt
- in note, 162

Browning, Elizabeth Barrett, 41-42, 184-185
- "Sonnet XLIII," 42, 184-185

C

Cantor Dust, 143, 146-147
- "The Cantor Dust," 105, 156-157

Cantor, Georg, 25, 141, 143, 146-147, 178, 180
- $\aleph_0$, aleph naught, 25,178
- infinity, countable, 25, 178
- infinity, uncountable, 25, 178, 180

Carroll, Lewis (Charles Lutwidge Dodgson), 193
Catalan Numbers, 51-52
chaos and fractals, 155
- in note, 42
- in poem, 159

Chapman, Robin, 158-160
- "Escaping the Mandelbrot Set," 158-160

Charles, Dorthi, 103-104
- "Concrete Cat," 103-104, 127

Cherry, Kelly, 82-83
- "Sequence Sonnet," 82-83

Christensen, Inger, 60
- *Alphabet*, 60

Churchill, Winston, 79
"Circle Limit III," 120
- in poetry, 191-192

Cohen, Marion, 173-174
- "Fermat's Last Theorem Proven," 173-174
- in note, 173

Complex Numbers, 20-21, 147-150
Complex Plane, 147-150, 154
- plotting points in, 148

convergence, 24
Cope, Wendy, 188-189
- "Depression," 188-189

Coxeter, H. S. M. and M. C. Escher, 119
- (*see* hyperbolic geometry)

Crane, Stephen, 187-188
- "War is Kind," 187-188

Cuddin, J. A., 186, 187
Cummings, E.E., 80, 127-128, 130
- experiments with sonnet form, 80
- "lis/-ten," 127-128
- ") when what hugs stopping each than silent is," 80-81

cyclic number, 70
in note, 70

D
dactyl, 34-36
Daniel, Arnaut, origins of Sestina, 66
Dante, 60
The Divine Comedy, 60
Daubechies, Ingrid, proof and poetry, 169-170
De Beaulieu, Eustorg, 97
"Gloire à Dieu seul," 97
Der-Hovanessian, Diana, 157-158
"Fractals," 157-158
in note, 157
Dekker, Thomas, 92
The Honest Whore, 92
Devaney, Robert
in note 142
Dickinson, Emily, 186-187
in note, 162
"Success is Counted Sweetest," 187
dimension,
fractal concept, 136, 138, 141-143, 145-146
Hausdorff-Besicovitch dimension, 142-143
in geometry, 87-89, 176, 178
in poetry, 161, 165,
in note, 162
Diophantus, 171-172
Diophantine Equations, 171-172
Doplicher, Fabio, 130
"Asymmetry of the Universe," 130
Duckworth, George E., 60
Structural Patterns and Proportions in Vergil's Aeneid: A Study in Mathematical Composition, 60
controversy over findings, 60
Dürer, Albrecht, 28-29
in note, 30
"Melancholia I," 28-29
dynamical systems, 155
in poetry, 157

E
Eddington, Sir Arthur, 171
"The Clock no question makes of Fasts and Slows," 171
Edwards, A. W. F., 45
Pascal's Arithmetic Triangle, 45
Einstein, Albert, 171-172
Theory of Relativity, 171-172
Escher, M. C., 87, 117-120, 122, 191-192
"Circle Limit III," 120
in poetry, 191-192
Coxeter, H. S. M., relationship to hyperbolic geometry120,122
"Symmetry Drawing E63," 119
tessellations, 117-120
Eratosthenes, 193
Euclid, 61, 77, 157-158
geometry, 119-120, 192
Golden Ratio, 61-63
in poetry, 77, 157-158, 192
Euclidean Plane, 119, 192

F
Fadiman, Clifton, 39, 60
Fermat, Pierre de, 171-172
Fermat's Last Theorem, 171-174
Wiles, Andrew
Fermat's Spiral, 93
poetry challenge, 172-174
Cohen, Marion, 173-174
"Fermat's Last Theorem Proven," 173-174
Mazur, Barry, 172-173
"Sing Fermat," 173
Teitelbaum, Jeremy, 172
Fibonacci Sequence (Fibonacci Numbers), 45, 48-49, 52-61, 62, 66, 104
Golden Ratio, 61-62
in music, 59
in nature, 54-59
in poetry, 59-61, 104
in Virgil's *Aeneid*, 60
Leonardo of Pisa, 49, 53
Liber Abaci, 53
Lucas Numbers, 53
origins, 53
rabbit breeding problem, 53-54
figurative language, 183, 184-185, 188-189, 193
Flegg, Graham, 30
in note, 30
Numbers, Their History and Meaning, 30
fractals, 12, 13, 78, 131-167, 193
basic concepts, 133-138
definition, artistic, 133
definition, mathematical, 134-136
Cantor Set (Cantor Dust), 146-147, 156-157

fractals (continued)
chaos and fractals, 155
in note, 42
in poem, 159
dimension, 134, 136-138, 141, 142, 143, 145, 146, 161, 165, 176
Hausdorff-Besicovitch, 142-143
dynamical systems, 155
in poetry, 157
fractal geometry, 12, 13, 131, 133, 141-155, 158-160, 162, 163, 167
in note, 163
Hilbert Curve, 142-143
history, 141-155
in nature, 131-132, 137-140
in poetry, 156-167
images in words, 156-157
Mandelbrot Set and language, 157-160
reading poetry with fractals in mind, 162-166
iterative (recursive) process, 52, 53, 134, 135-136, 141, 143, 145, 146, 147-149, 150, 155
in poetry, 160, 161, 164, 166
Julia Sets, 147-149, 154-155
Koch Snowflake, 143-145
link between math and poetry, 167
Mandelbrot, Benoit, 141, 143, 167
in note, 157
in poetry, 156-158
Mandelbrot Set, 132, 141, 147-155
connection to Julia Sets, 154-155
in poetry, 156, 157-160, 166
measure theory, 143
Hausdorff, Felix, 143
Besicovitch, Abram, 143
Peano Curve, 141-142
scale, 131, 134-136, 138-140, 147
in poetry, 158-160, 164, 166
self-similarity, 134, 136, 138-139, 142, 147, 151-152
in poetry, 158-161, 164-166
Sierpinski Triangle (Sieve), 145-146
space-filling curves, 141, 163
"fractal dust" in poetry of Wallace Stevens
in note, 163
free verse, 34, 42
in note, 162
Fulton, Alice, defining fractal poetry, 162

G

Gardner, Martin
in note, 70
Garrison, Denis, 104-105
Fibonacci Sequence in poetry, 104
"Nautilus," 105
geometry (*see* Euclid, Euclidean Plane, fractal geometry)
Golden Ratio (Divine Proportion, Golden Mean, Golden Number, Golden Section, Φ, phi), 12, 45, 53, 61-65, 185
definition, 61
Golden Rectangle, 62-64
Golden Spiral (Equiangular Spiral, Logarithmic Spiral), 64-65, 93
in art and architecture, 62-63
in poetry, 65
limit of Fibonacci Ratios, 61-62
Gomringer, Eugen, 102
concrete poetry "constellations," 102
untitled poem, 102

H

Hancock, Marga Rose, 184
"Reflections on Poetry + Proof," 184
Hausdorff-Besicovitch dimension, 142-143
Hein, Piet, 193
Herbert, George, 99-100, 127
"Easter Wings," 100
Herz-Fischler, Roger, 60
Higgins, Dick, 163
in note, 97
Hilbert, David, 141-142, 177
Hilbert Curve, 142-143
and poetry, 162-163
Hilbert Hotel, 177-178
Hofstadter, Douglas on M. C. Escher, 122
Holmes, Janet, 156
"The Pattern," 156
Holmes, Oliver Wendell, 65
"The Chambered Nautilus," 65
Hughes, Langston, 42-43
"The Negro Speaks of Rivers," 42-43
Hui, Yang, 45
hyperbolic geometry, 119-121

I

i, $\sqrt{-1}$, 20, 148
iamb, 34-36

infinity, 24-26, 169, 178-180
  C, Continuum, 178
  cardinality, 178
  in poetry, 180, 189-192
    Blake, William, 180-181
    Kerouac, Jack, 189
    Leopardi, Giacomo, 190-191
    Shakespeare, William, 190
    Tagore, Rabindranath, 189
    Wordsworth, William, 190
  infinity, countable, 25-26, 178
  infinity, uncountable, 25, 178-180
Integers, 17
irony, 182, 188
Irrational Numbers, 19-21, 24, 25, 61, 170, 178
iterative (recursive) process, 52, 53, 134, 135-136, 141, 143, 145, 146, 147-149, 150, 155
  in poetry, 160, 161, 164, 166

J

James, Henry, 182
Johnson, James Weldon, 123-124, 125, 127
  "Sence You Went Away," 123
Johnson, Michael L., 60-61, 191-192
  "Fibonacci Time Lines," 60-61
  "M. C. Escher's *Circle Limit III,*" 191-192
Jordon, Barbara, 194
  "This Poem," 194
Julia, Gaston, 147
  Julia Sets, 147-149, 154-155

K

Kaufman, Gerald Lynton, 90, 95
  "Sigh-Angles," 90
  "Spiralay," 95
Keats, John, 183-184
  "Ode on a Grecian Urn," 184
Kenney, Richard, 170
Kerouac, Jack, 189
  "The Moon," 189
Khayyám, Omar, 45, 193
Koch, Helge von, 141, 143, 145, 147
  Koch Snowflake, 143-145, 176
Koshy, Thomas and George E. Duckworth, 60

L

labyrinths, puzzles, 96-97
  SATOR word square, 96
  "Gloire à Dieu seul," 97
Lake, Paul (in note), 162
Lear, Edward, 38
Leonardo of Pisa (*see* Fibonacci)
Leopardi, Giacomo, 190-191
  "The Infinite," 190-191
Leibnitz, Gottfried, 25
limerick, 38-41, 59-60, 172
limit, 24, 62
literary terms, 182-183, 186
Livio, Mario, 59
Lucas Numbers, 49, 53

M

magic square, 28-30, 96, 193
  Chinese relic, 30
  Dürer, Albrecht, 28-29
Mandelbrot, Benoit, 141, 143, 158, 167
  The Fractal Geometry of Nature, 167
  Les Objets Fractals, 141
  mathematicians who inspired, 141-145
  Mandelbrot Set, 132, 141, 147-155
    connection to Julia Sets, 154-155
    in poetry, 156, 157-160, 166
Maor, Eli, 25, 26
math games (*see* number puzzles, labyrinths)
mathematical proof (*see* proof)
Mazur, Barry, 172-173
  in note, 173
  "Sing Fermat," 173
McKay, Claude, 79-80, 82
  "If We Must Die," 79-80
"Melancholia I," 28-29
metaphor, 10, 41, 65, 87, 123, 129-130, 157, 158, 160, 163-164, 183, 184, 189, 190
meter, 12, 31-36, 38, 41-42, 66, 76, 81, 98, 164
  accentual verse, 32-33
    accentual-syllabic verse, 33-34
  syllabic verse, 33-34, 60
metrical feet, 34-35
Millay, Edna St. Vincent, 37-38, 77-78, 124-127
  "Euclid alone has looked on Beauty bare," 77
  "Recuerdo," 37
  "Small Hands, Relinquish All," 124-127

Morse Code in pattern poetry, 102-103
Mother Goose, 26-27, 31-33

N

Nicolson, Marjorie Hope, 190
North, Charles, 81-82
  "Sonnet," 81-82
numbers, 10, 15-30, 32, 61, 70, 134, 136, 145-150, 170-172, 174, 178-179
  number sequences, 21-23
  number systems, 16-21
    Integers, 17-18, 19, 21-22, 25-26 53, 134, 136, 145, 171-172, 178-179
    Numbers, Complex, 20-21, 147-148, 149-150
    Numbers, Irrational, 19-21, 24, 25, 61, 170, 178
    Numbers, Natural, 16-17, 21-22, 25-26, 28, 32, 70, 174, 178-179
    Numbers, Negative, 17
    Numbers, Rational, 18-20, 25, 178
    Numbers, Real, 20, 25, 178-179
    Numbers, Whole, 17, 21, 22
  number puzzles, riddles, 26-30
nursery rhymes, 11, 31-33, 36, 38
Newton, Isaac, 25

O

Oliver, Mary, 128-129
  "Something," 128-129
  symmetry and organic form, 128-129

P

paradox, 12-13, 169, 176-178, 181, 186-189, 193
  Hilbert's Hotel, 177-178
  in poetry, 181, 186-189, 193
  mathematical types, 176
  Russell, Bertrand and Barber's Paradox, 177
  Zeno's Paradox (*see* Zeno)
Pascal, Blaise, 45, 193
  Traite du Triangle Arithmetique
Pascal's Triangle, 12, 45-52
  binomial expansion, 47
  Catalan Numbers, 51-52
  combinations, 50
  sum of diagonals, 48
  Natural Numbers, 49
  powers of 11, 50
  Triangular Numbers, 51
Passerat, Jean and origins of villanelle, 83
pathological curves, 143, 163, 176
pattern poetry (shaped poetry, concrete poetry), 96-105, 156
  labyrinths, puzzles, 96-97
    SATOR word square, 96
    "Gloire à Dieu seul," 97
  geometric forms and guidelines, 97-100
  in note, 97
Peano, Giuseppe,141-142, 145, 148, 176
  Peano Curve141-142, 145, 163, 176
permutation, 68-70, 71-76
  cyclic numbers, 70
  in sestina, 68-69, 71-76
Petrarch, Francesco 76-78
  origins of the sonnet, 76
  "Zephyr returns . . . ," 76-77
Φ, phi (*see* Golden Ratio)
π, pi, 24-25
  Viète, François, 24-25
plane figures, 88-92
Pollard-Gott, Lucy, 163
  in note on fractal patterns in work of Wallace Stevens, 163
proof, 90, 169-176, 178, 179, 181, 184-186, 193
  contradiction, proof by, 174
  direct proof, 174
  exhaustion, proof by, 174
  fallacy, 175-177
  in poetry, 181, 184-186, 193
    commemorating mathematics, 171, 172-174
    Fermat's Last Theorem poetry challenge, 172-174
  induction, proof by, 174
  mathematical proof, 169-176
    corollary, 170
    lemma, 170
    theorem, 170
prosody, 34-36
Puttenham, George, 97-99, 104
  The Arte of English Poesie, 97-98
  guidelines for shaped poetry, 97-98
  "Her majestie, for many parts in her," 99
puzzles (*see* number puzzles, labyrinths)
Pythagoras, 19
  Pythagorean Theorem, 19, 171

Q
Quiller-Couch, Arthur T., 40-41
"A New Ballad of Sir Patrick Spens," 41

R
Rational Numbers, 18-19, 25, 178
Real Numbers, 20, 25, 178-179
rhyme, 10, 12, 31-34, 36-39, 66, 70-71, 76-78, 81-84, 95, 98, 106, 123-126, 164
rhythm, 31-35, 43
Robinson, Edwin Arlington, 83-84
"The House on the Hill," 83-84
Rogers, Pattiann, 160-161
"Fractal: Repetition of Form over a Variety of Scales," 160-161
Russell, Bertrand, Barber's Paradox, 177

S
Sakaki, Nanao, 164-166
"A Love Letter," 165-166
Sandburg, Carl, 15
"Arithmetic," 15
SATOR word square, 96
sestina, 66-76
Bishop, Elizabeth, 66-68
cyclic number, 70
double sestina, 71-76
permutation, 69-70, 71, 75
Shakespeare, William, 78-79, 82, 163-164, 190
"Sonnet 73," 78-79, 163-164
The History of Troilus and Cressida, 190
shaped poetry (*see* pattern poetry)
Shattuck, Roger, 101
"It's Raining" ("Il Pleut"), 101
Sierpinski, Waclaw, 141, 143, 145-146
Sierpinski Triangle, 145-146
Silliman, Ron, 60
Tjanting, 60
Siqueira, Rodrigo, 105, 156-157
"Cantor Dust," 157
fractals, 156-157
Solt, Mary Ellen, 81, 102, 102-103, 105
"Forsythia," 102-103
"Moon shot," 81, 105
sonnet, 12, 38, 41, 66, 76-85, 105-106, 157-158, 163-164, 184-185
couplet, 78, 82, 164
octave, 76, 78, 81
quatrain, 78, 83, 164
sestet, 76, 78, 81
sonnet, English (Shakespearean), 76, 78-79
sonnet, Italian (Petrarchan), 76, 77-78
spiral, 12, 56-59, 61, 93-95, 185, 188
Archimedes', 93
Fermat's, 93
Golden (Logarithmic, Equiangular), 64-65, 93
how to construct, 64
in nature, 56-59, 94
in poetry, 95, 188
hyperbolic, 93
spondee, 34-36
Stern, Gerald, 80
Swinburne, Algernon Charles, 71-76
"The Complaint of Lisa," 71-76
syllables, 32-39, 60, 104
syllabic verse, 33-34, 60
syllables, stressed, 32, 34, 37, 39
syllables, unstressed, 32, 34
symbols in poetry, 80-81, 102-103, 105, 182
symmetry, 87, 106-130
basic symmetries106-109
glide reflection, 108, 114-116
reflection (bilateral), 106-109, 112-116
rotation, 106-109, 110-112
translation, 106, 108-109, 114-116, 124, 125
in architecture, 113
in nature, 113-114
in poetry, 123-130
invariant, 106-107
isometry, 109, 114-115
line symmetries (frieze patterns), 114-115
n-fold, 110-112
plane symmetries (wallpaper groups), 109, 116-117
tessellations, 116-122
"Symmetry Drawing E63," 119

T
Tagore, Rabindranath, 189
"On the Seashore," 189
Taylor, Richard (*see* Wiles)
Teitelbaum, Jeremy and Fermat's Last Theorem poetry challenge, 172
tessellations, 116-122

tessellations (continued)
  Euclidean geometry, 116-119
  hyperbolic geometry, 119-122
    H. M. S. Coxeter, 119
  M. C. Escher (*see* Escher, M. C.)
Triangular Numbers, 51
trochee, 34-36

V

Viète, François, 24-25
  approximation to π, 25
villanelle, 12, 66, 83-85
Virgil (Vergil), 60
  Aeneid and George E. Duckworth, 60
von Koch (*see* Koch)

W

Weierstrass, Karl, 169
Whitman, Walt, 186
  "When I Heard the Learn'd Astronomer," 186
Wiles, Andrew, 172-174
  Fermat's Last Theorem, 172
  Taylor, Richard, 172
  poetry challenge, 172
Wilson, Robert M., 60
  "Fibonacci Memory," 60
Wiltz, Chris, 85
  "Icebreaker," 85
Wordsworth, William, 190
  "The Borderers," 190

Z

Zeno of Elea, 24
  Zeno's Paradox, 24, 176, 188-189
  in poetry, 188-189
zero, 16-17

Everglades Fractal

A ~~raindrop fell~~ Rain falls
~~and joined in~~ into a ~~temperate~~ shallow sea
Where bivalves siphoned sustenance

~~A raindrop fell~~

time
archaic
↓
present

Printed in the United States
203334BV00005B/316-333/P

9 789042 023703